W0042796

International Political Economy Series

General Editor: **Timothy M. Shaw**, Professor of Political Science and International Development Studies, Dalhousie University, Halifax, Nova Scotia

Titles include:

Leslie Elliott Armijo (*editor*)
FINANCIAL GLOBALIZATION AND DEMOCRACY IN EMERGING MARKETS

Robert Boardman
THE POLITICAL ECONOMY OF NATURE
Environmental Debates and the Social Sciences

Gordon Crawford
FOREIGN AID AND POLITICAL REFORM
A Comparative Analysis of Democracy Assistance and Political Conditionality

Matt Davies
INTERNATIONAL POLITICAL ECONOMY AND MASS COMMUNICATION IN CHILE
National Intellectuals and Transnational Hegemony

Martin Doornbos
INSTITUTIONALIZING DEVELOPMENT POLICIES AND RESOURCE STRATEGIES IN EASTERN AFRICA AND INDIA
Developing Winners and Losers

Fred P. Gale
THE TROPICAL TIMBER TRADE REGIME

Keith M. Henderson and O. P. Dwivedi (*editors*)
BUREAUCRACY AND THE ALTERNATIVES IN WORLD PERSPECTIVES

Jomo K.S. and Shyamala Nagaraj (*editors*)
GLOBALIZATION VERSUS DEVELOPMENT

Angela W. Little
LABOURING TO LEARN
Towards a Political Economy of Plantations, People and Education in Sri Lanka

John Loxley (*editor*)
INTERDEPENDENCE, DISEQUILIBRIUM AND GROWTH
Reflections on the Political Economy of North–South Relations at the Turn of the Century

Don D. Marshall
CARIBBEAN POLITICAL ECONOMY AT THE CROSSROADS
NAFTA and Regional Developmentalism

Susan M. McMillan
FOREIGN DIRECT INVESTMENT IN THREE REGIONS OF THE SOUTH AT THE END OF THE TWENTIETH CENTURY

James H. Mittelman and Mustapha Pasha (*editors*)
OUT FROM UNDERDEVELOPMENT
Prospects for the Third World (second edition)

Lars Rudebeck, Olle Törnquist and Virgilio Rojas (*editors*)
DEMOCRATIZATION IN THE THIRD WORLD
Concrete Cases in Comparative and Theoretical Perspective

Howard Stein (*editor*)
ASIAN INDUSTRIALIZATION AND AFRICA
Studies in Policy Alternatives to Structural Adjustment

International Political Economy Series
Series Standing Order ISBN 978-0-333-71708-0 hardcover
(*outside North America only*)

You can receive future titles in this series as they are published by placing a standing order. Please contact your bookseller or, in case of difficulty, write to us at the address below with your name and address, the title of the series and the ISBN quoted above.

Customer Services Department, Macmillan Distribution Ltd, Houndmills, Basingstoke, Hampshire RG21 6XS, England

The Political Economy of Nature

Environmental Debates and the Social Sciences

Robert Boardman
McCulloch Professor of Political Science
Dalhousie University
Nova Scotia
Canada

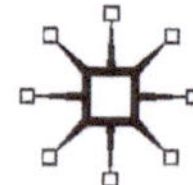

© Robert Boardman 2001

All rights reserved. No reproduction, copy or transmission of this publication may be made without written permission.

No paragraph of this publication may be reproduced, copied or transmitted save with written permission or in accordance with the provisions of the Copyright, Designs and Patents Act 1988, or under the terms of any licence permitting limited copying issued by the Copyright Licensing Agency, 90 Tottenham Court Road, London W1P 0LP.

Any person who does any unauthorised act in relation to this publication may be liable to criminal prosecution and civil claims for damages.

The author has asserted his right to be identified as the author of this work in accordance with the Copyright, Designs and Patents Act 1988.

First published 2001 by
PALGRAVE
Houndmills, Basingstoke, Hampshire RG21 6XS and
175 Fifth Avenue, New York, N. Y. 10010
Companies and representatives throughout the world

PALGRAVE is the new global academic imprint of St. Martin's Press LLC Scholarly and Reference Division and Palgrave Publishers Ltd (formerly Macmillan Press Ltd).

ISBN 978-0-333-80015-7

This book is printed on paper suitable for recycling and made from fully managed and sustained forest sources.

A catalogue record for this book is available from the British Library.

Library of Congress Cataloging-in-Publication Data
Boardman, Robert, 1945–
The political economy of nature : environmental debates and the social sciences / Robert Boardman.
p. cm.
Includes bibliographical references and index.
ISBN 978-0-333-80015-7
1. Environmental policy—International cooperation.
2. Environmental sciences—Social aspects. I. Title.
GE170 .B63 2001
363.7'0526—dc21

00–067096

10 9 8 7 6 5 4 3 2 1
10 09 08 07 06 05 04 03 02 01

For Christine

Contents

Acknowledgements

I have benefited in innumerable ways from students in my classes on global environmental politics and on international relations theory. I also learned much in exchanges with former doctoral students, now graduated: Luc Ashworth, on functionalist theories in international relations; Robert Webber, on post-structuralist theories and identity politics; Karen Beazley, on landscape ecology and endangered species; and Fahim Quadir, on globalization and civil society in the South. My colleague Florian Bail was an ever-helpful and authoritative voice on continental political theory. If I had the ability to reflect competently on these insights, this would have been a much better book.

Hokkaido was a perfect setting in the autumn of 1999 for thinking through some problems. I would especially like to thank Dean Tadashi Okudaira of Sapporo International University for his kind hospitality in making this possible.

The Social Sciences and Humanities Research Council of Canada funded research as part of a larger programme on international environmental politics. I gratefully acknowledge this support. I would also like to thank my own university for a sabbatical leave research grant.

A part of Chapter 2 is a shortened and revised version of an article in *Global Society*, **11** (1), in 1997; and a section of Chapter 7 is adapted from parts of my chapter in *New Perspectives on International Functionalism*, edited by Lucian Ashworth and David Long and published in the International Political Economy series in 1999.

Chester Basin R. B.

Part I
Echoes

1 Infused Opinions

Discourses on the relations between societies and their natural environments have been a persisting feature of the politics of modernity. Writers have routinely brought in such topics in the course of elaborating general social and political theories. Engels pondered the significance of the divorce of capitalist societies from nature. The forms taken by human polities, Montesquieu thought earlier, and the cultural traits of individuals in them, are due to differences in the climatic conditions of societies. As more recognizably 'environmental' questions began to structure debates, focal points proliferated. In the late nineteenth century they touched on national parks, the millinery trade and problems of wild birds as hazards to agriculture. The global balance of population and natural resources loomed menacingly in the 1940s and again in the 1970s. Sick buildings in the 1990s, supersonic aircraft for a time in the 1960s, migratory species, the capacities of Southern economies to finance environmental reform, agrochemical use by golf courses and suburban households, changing weather patterns: there has been no shortage of grist for the environmentalist issue-mill.

Theory, politics and voice

The contention that these kinds of questions are puzzles in larger theoretical enterprises has also lasted, in part because of the unavoidably political character of environmental problems and the proliferation of contexts in which these have been situated. Global environmental discourses are theoretically relevant, first, because they contain cognitive maps that can be used to investigate social and ecological worlds, and, second, because special tools are often required to understand them and their associated practices.

Theory, though, has its own array of contested meanings. One theoretical perspective common to, or hidden beneath, much environmental discourse suspects that attention to theory is a distraction. We do not need to reflect on social theory before setting up a recycling scheme or finding ways to prevent people from drenching their lawns with pesticides. Real problems exist, that is, and actions are needed to deal with the ecological crisis. The function of discussion or writing, in this view, is to help push the process of social change forward by clarifying the nature of problems, effecting a better translation of science into policy or action by civil society groups, and raising public awareness. In its postmodern guises, theory often looks more like a threat than an aid to the goals of environmentalism (Coates, 1998: 184–5). The task, especially given the uncertainty of much of the science related to key environmental questions, is to work pragmatically towards more immediate social-action or public-policy goals (Light and Katz, 1996: 1–2; Farber and Juma, 1999).

There are other grounds than a pragmatic hankering for results for resisting the charms of theory. Environmentalist questions do not seem to outsiders to generate theoretical questions. They form an isolatable clump of varying, but usually fairly marginal, significance in the course of election campaigns or water-cooler chats. Sympathetic critics noticed that during the 1990s environmental questions lost some of their former capacity to mobilize political activism (Tranter, 1999: 348–50). Environmental topics have also had a variable history in social theorizing. They have often been forced to retreat by more self-evidently emancipatory or justice-based commentaries. Attempts through situationist-aesthetic means to revitalize the Marxist tradition, for example, have focused on problems of consumerism, culture and the media, but have tended to show little interest in issues pressed by ecological critics (Best and Kellner, 1999: 130–2). Claims that environmentalism has a solid base in theory, and a capacity to enliven wider social discourses, thus look precious. Such appeals become more comprehensible for critics when dismissed either as a typical legitimizing strategy of single-issue groups, or as reflections of the private self-development paths navigated by their members.

Similar complaints, though from a different definition of theory, occur to the more empiricist of social scientists. Theory is 'simply a research tool, and not the end-product of research' (Castells, 2000a: 6). Theorizing helps usher in the initial sense of order that is the starting point for empirical research in a complex world (Rosenau and Durfee, 1995: 2). The question is linked to definitions of the appropriate methodologies of environmental enquiry. Critics from the natural sciences have frequently maintained that since data collection and analysis hold the keys to effective environmental policies, it is the natural sciences that constitute the primary foundation of environmental knowledge. Despite this chronic cultural bias, there has been increasing recognition in these disciplines of the importance of social and economic processes. 'It has been clear from the beginning that resource management problems are human ones rather than technical ones. Nevertheless the vast bulk of work in the field has been narrowly and technically focused. Our magical expectation that the technical detail will somehow be integrated into effective policy has not been fulfilled' (Ludwig, 1994: 112). Social scientists have nonetheless nursed long-standing fears that their own contributions tend to be quickly marginalized in the larger interdisciplinary context of environmental policy debates. If human actions cause environmental problems, they insist, then enquiry should logically begin in the grand project of the humanities and social sciences.

Theory is inescapable in part because politics is implicated in environmental debates. Environmental discourses, in Greenpeace or in scholarly writings, are above all about *doing something*. They are about the actions required by civil-society groups, government agencies, international organizations, or by individuals in their waste-disposal, energy-consumption, dietary and other daily habits, if environmental problems are to be solved. A recurrent issue in debates on global climate change, for example, has been on the significance of anthropogenic factors such as greenhouse-gas (GHG) emissions. Naturally occurring variations in climate fall outside the boundaries of 'environmental' discourses, or, more accurately, enter these only as a means of clarifying the parameters of anthropogenic change. Embedded constructions of the 'political' likewise structure environmental discourses and activities. In a pluralist setting, politics can be environmentalist shorthand for the dark forces that stand in the way of change. Or it can be embraced as the least worse guarantor of practical results in an imperfect world. Writing of the USA, Smith refers to the 'paradox of environmental policy' that arises from neglect or misunderstanding of politics and political processes. We 'often understand what the best short- and long-term solutions to environmental problems are; yet the task of implementing these solutions is either left undone or is completed too late' (1995: xi).

Firstly, there are conflicts over how to achieve environmental policy goals. These disputes, ostensibly over means rather than ends, are often bitterly contested. As the Canada–USA Pacific salmon conflict of the late 1990s demonstrated, shared conservation cultures built into resource regimes can quickly collapse under the pressure of competing interests. Broad anti-pollution goals tend to be accepted within OECD countries, but debates are characterized by sharp disagreements on the choice of policy instruments. In public-choice terms, agreement on social goals does not deter defectors. An individual can avoid the costs of car emission-control repair while freeloading on the cleaner-air benefits of regulation and the compliance of others. Green areas in malls or campuses are vulnerable to the choices of pedestrians taking short cuts across grass. Multilateral negotiations on the ozone layer and climate change have been marked by protracted North–South and intra-West conflicts over the politics of means, including timetables, market versus regulatory instruments, exemptions and a variety of side-payment issues.

These considerations overlap, secondly, with clashes over goals and process. Even recycling schemes and local community pesticide-use regulation turn on traditional issues of individual rights, obligations and the nature of democratic governance. They are not simple problem-solution technical questions. Many underlying issues raised in environmental policy debates, by supporters and critics alike, are inextricably connected with old questions of the good life and the just society. They include problems of land ownership and conceptions of property and commons rights, the nature of political obligation, definitions of community, trade-offs between market forces and social justice, the political legitimacy of international organizations in a world of sovereign states, and the place of civil disobedience in democratic societies. Environmental questions have become central to the standard discourses of Western polities, and the professionals handling them part of the conventional state apparatus. The traditional lefts and rights of Western liberal democracies have each taken on board some aspects of environmentalist thought and practice. But each has kept some prejudices alive: the respective suspicions that environmentalism too easily neglects either distributive justice and the correcting of inequalities, or property rights and the creation of wealth. Societies have responded in differing ways to the problems generated by environmental conflicts. Cooperation among environmental NGOs, universities and potato-farmers in Wisconsin has helped to promote state agriculture while containing problems such as pesticide spray drift. Some conflicts have defied short-term resolution. Reinvigorated claims to hunting and fishing rights by aboriginal organizations erupted into a major political and constitutional dispute in the Canadian Maritime province in the late 1990s.

Thirdly, environmental hazards and benefits are unevenly distributed. Poor and marginalized groups in any society are affected more seriously by geographical variations in air and water pollution. Support for environmentalism is a more comfortable option for the affluent. Some environmental beliefs and practices have become markers of membership of richer groups in Western countries. These groups are better placed to resist burdens from the selection of toxic waste sites, and are usually immune to the job-loss consequences of forestry or other resource-sector environmental regulation. Parks and protected areas have eliminated or reduced options for common-use fishing, hunting or grazing in colonial and postcolonial Africa, as well as in the rural hinterlands of the Western industrialized world.

A fourth sense of politics stems from the meta-choices made by societies in their relations with natural environments. Choice in the context of these ecological contracts is a contestable term. In practice societies rarely if ever make such very large-scale decisions. The unintended or unanticipated ecological consequences of even the simplest decisions make environments inherently uncontrollable (and the visionary element required to make this kind of social engineering hypothetically feasible would probably at the same time render it democratically undesirable). It is useful nonetheless to retain at least the metaphor of societal choice. This highlights the conventions of the imagined constitutions that govern the relations between societies and their environments. Such deals are partial; for example, non-human species do not participate in their making, or are at best dubiously represented through the voices of others.

Finally, nature is a resource for political ideologies and the ideas that structure societies. It figures in constructs as diverse as the conceptions of countryside that have traditionally nourished conservative definitions of 'natural' communities; the formally articulated and promoted love-of-nature ethic that was embedded in the ideological cultures of the former socialist states of Eastern Europe; the social movements of the USA that have drawn variously on Thoreauvian self-development ethics and community-based progressivism; and even in some of the roots of National Socialism in Germany in the 1930s, by way of an earlier idealization of the land, particularly forests, and its role in the defining of peoples. The natural environment was an indispensable context of many nineteenth-century European nationalisms. Its features and locales echo through music and the arts, as Smetana, for example, traces the course of the river Vltava in the second section of *Má Vlast*, or as Chekhov sketches Astrov's project for a reforested and ecologically renewed Russia in Act 1 of *Uncle Vanya*. Representations of the wilderness that confronted pioneers and settlers were potent elements in the construction of New World nationhoods.

Ideological tensions of different kinds continue to affect contemporary environmental politics. Rural–urban cleavages are a source of some. Moose or deer in eastern Canada, or small migratory birds in parts of Italy and France, may be seen either as moving protein and a staple of community life and family cultures, or as sources of poetic wonder. Some city folk yawn at both responses. European and North American environmental movements have been shaped by versions of the realist-fundamentalist split originally identified with the German Greens. Evaluations of the political acceptability of working with, or within, governments vary greatly. The predicament is accentuated as some environmental NGOs, particularly in the USA, have themselves grown in wealth, political clout and global reach. Globalization complicates the issue further as some groups – as in the Seattle anti-WTO and the Washington anti-IFI (international financial institution) protests of 1999 and 2000 – come to see governments as conflicted instruments of larger economic and political structures. Ideological conflicts over environmental questions in the South arise when these are evaluated in the context of economic development goals, or, as in Malaysia and Indonesia during the 1980s and 1990s, when environmental debates become intertwined with the discourses of human rights and democratization.

This brings us back to the question of theory. The term signifies different things among the disciplines of the humanities and the social sciences, and between scholarly discourses and those outside academia. The traditional crafts of historians were threatened by definitions of theory associated with sociology in the 1960s, and by others stemming from the practices of post-structuralist critics in the 1980s (Hunt, 1990: 95–6). Many such questions evaporate in the natural sciences. Theory is simply 'the "text" in terms of which the model is specified, instructions are given on problem-solution, and so forth' (Weinert, 1999: 304). Many questions of environmental enquiry straddling the social as well as the natural sciences are indeed best approached as empirical ones. Quantification of the relations and causal links sketched out in models of global ecological change is an urgent requirement. Other questions, including both analytical and normative problems, are not so easily despatched. Putting states at the centre of global environmental enquiry is a theoretical strategy, though it points towards and is reinforced by empirical research. Empirical research can help fill out or redefine, but in itself cannot answer, questions about how best to trade off jobs and pollution, or whether tax revenues should be put into saving endangered species of wildlife.

In environmental debates there are occasionally traces of theory viewed as a good thing. Theory may be implicitly defined as the something-useful that can result from reflection and experimentation, or from standing back from immediate problems. As Mr Brooke says in *Middlemarch*, 'I have always been in favour of a little theory: we must have Thought; else we shall be landed back in the dark ages.' But it can be taken too far – 'over the hedge', as he puts it. Standing back to see the bigger picture is not good advice for someone perched on a ladder. Except in the restricted sense of provisionally tested or testable propositions, theory tends to remain a blind alley for rugged social-science empiricists. It is also a puzzling diversion in exchanges between academics and practitioners. Researching his realist analysis of foreign policy in the early 1990s, George had to disguise the word: 'I quickly found that the eyes of policy specialists glazed as soon as I used the word "theory". But they nodded approvingly when I spoke of the need for "generic knowledge"' (1999: 91). For other audiences, theory becomes a code-word for a suspect postmodernism. Yet the term still has an aura Wildavsky discovered when he was a graduate student: 'In political science was the word, and the word was theory, pronounced reverently, growing larger with the telling . . . Always alluring, forever, except for the favored few, unreachable' (1989: 28).

It is the latter sense of theory, though not the project of theory itself, that has been increasingly challenged in the humanities and social sciences. The critique is dual, resting on the opposed foundations roughly characterizable as those of subjectivity and objectivity, each self-defined as 'critical'. In the first, theory in the conventional senses in which the word is used in the social sciences is found wanting because of its epistemological and ontological assumptions of a knowable world; its faith in the existence and capacities of a static, removed and observing subject, and its lack of appreciation of the significance of people as meaning-producers; and its optimistic, possibly dangerous, belief in the feasibility of the juxtaposition of these beliefs producing generalizable law-like observations. In the second, theory is found to be unavoidably tied in varying ways to economic and social conditions and the ambitions of actors. It is accordingly bad or good, in terms of knowledge accumulation and its social consequences, depending on such things as where its producers are located in social structures, what their normative evaluations of these structures are, and the degree to which actors reflexively embrace theory as an engine of social change.

This portrait is overly simple, but it accentuates some of the key issues we encounter when studying environmental discourse and practice. The voice of the observer of these is one of the more problematic. Attention to theory sharpens the writer's predicament by compelling choice among various forms of detachment, apparent objectivity, empathic engagement or participant observation. Bramwell met indignant opposition from some environmentalists when trying to study them. Her job, they implied, was to look out into the real world of environmental crisis, not to analyse the people solving the problems (1994: ix). A tilt towards partisanship avoids this response. In his analysis of environmental NGOs, Wapner (1996) writes from a perspective that shares much of the environmentalist's self-image of a fighter against government hypocrisy and the ecologically destructive greed of multinational corporations. Some environmental writing is an offshoot of action, produced by actors or their sympathizers, and viewed within the environmental world as a key element in the larger praxiological task. The first of these strategies, taken too far, risks loss of the hermeneutic understanding that comes from making contact with environmentalists' thoughts and feelings and those of their critics. It reaches back for the lunar, later the Martian, observer – a sceptical, distant and occasionally querulous questioner of unfamiliar conventional wisdoms – who haunted European scholarly discourses in the nineteenth century. The second option misses opportunities for social and political analysis by taking the stated beliefs of sympathizers at face value, and by ignoring both interest as a factor in the political lives of groups and individuals, and power as an element in their relations with others.

Much writing on environmental topics tends in practice towards the voice of the accomplice. It is 'mundane', in Pollner's use of the term to describe the predicament of sociologists, in the sense that it tends to share much of the vocabulary, narrative style, values, definitions of problems or assumptions of those being observed or written about. The task of the critical observer is one of defamiliarization of the vocabularies and practices of the relatively well-known environmentalist world. This involves a creative interplay among participants' and observers' views. Sociology, with its ingrained optimism about its capacity to reveal the unexpected disguised in the familiar (Portes, 2000) and to 'crush the walls of the obvious and self-evident' (Bauman, 2000: 79), is a useful guide on this trek.

Detachment, though, like its occasional drinking companion playfulness, is an unwelcome intruder at insiders' parties. Sometimes seen fiddling away like Nero, this forlorn virtue, indicted elsewhere as an accomplice in practices of cultural oppression, appears almost criminally oblivious to humanity's laddish disregard of its *oikos*. Yet at least a partial rehabilitation is warranted if we consider the sheer diversity of interpretations of the environmental. Regarbed as irony, detachment becomes 'a way to live or write or read at a distance from multiple and otherwise irreconcilable positions without directly engaging these positions' (Livingston, 1997: 84). But it is not reducible to a posture of moral indifference. It has richer ethical tones, and respects the autonomy and integrity of the multiple sets of beliefs it investigates. As Marx might perhaps be persuaded to say, if we were to turn him, in turn, on his head: environmentalists have only changed the world in various ways; the point is to interpret it.

A tale of two *oikoi*

The observer-writer cannot be somehow magically removed from this world of multiple interpretations. We can nonetheless make some pragmatic as-if assumptions of epistemology and ontology. As Ward writes in his study of theology and critical theory, 'What standpoint can be taken above discourse from which to survey and plot the lines of social force which relate intellectual loci? No such standpoint exists, except a pragmatic one. That is, we have to write *as if* such a standpoint were possible, *as if* our own discourse belonged to none of the intellectual disciplines upon which it is commentating, *as if* its standpoint was neutral and its author omniscient' (1996: 2).

The account in subsequent chapters centres on the structuration of ecological discourses and practices, particularly those that aim to situate globality in their frameworks. In this chapter and chapter 2 I argue that environmental arguments contain or imply a variety of theoretical and disciplinary perspectives. Although in many ways environmental enquiry has become increasingly and self-consciously interdisciplinary, this process has been less evident the closer we move to the theoretical cores of disciplines or of grand transdisciplinary movements. Two of these – political economy and ecology – nonetheless have a continuing and unavoidable presence in the study of the issues in environmental discourses. They point towards differing frameworks for understanding global environmental problems.

Several competing theoretical sources of global ecological insights are investigated in chapters 3–7. These are each linked to substantive research questions, but more importantly, I argue, they constitute alternative theoretical approaches to understanding the linkages among economies, societies and ecosystems. They have differing significance too in terms of theoretical interactions with either ecologically or economically grounded frameworks. They contribute both to general-theory development centred on ecological enquiry, and also to discussions of the ways global-society and political-economy frameworks can be enhanced by making environmental issues more central. These theoretical bases are respectively (1) those of the earth-system sciences and the links between these and developments in ecology (chapter 3); (2) subjectivity and the problematic of representations of the environmental domain (chapter 4); (3) approaches to global ecological issues that draw insights from utilitarian and other ethical traditions (chapter 5); (4) methodological individualism, in particular the alternately diverging and overlapping paths created by social-theory conceptions of agency on the one hand, and the perspective of rational expectations and choice on the other (chapter 6); and, finally, (5) applications to environmental enquiry of notions of governance, particularly at the global but also at domestic and sub-national levels (chapter 7).

This sequence of topics facilitates global-ecological investigation at the interfaces of several traditionally central dualities of social thought: that is, those centred on nature–humanity, subjective–objective, normative–positive, individual–collective, and government–society distinctions. In addition, the diverse themes of environmental discourses – which are the focus of part of the discussion in chapter 4 – also raise, even if only implicitly, many of the issues contained within these larger constructs and thus themselves become a source of theory-building activity.

The discussion in chapter 8 focuses on the integrative possibilities arising from this account. The issue is also explored in various ways in earlier chapters. Each of the broad approaches explored in chapters 3–7 contains mechanisms for enclosing or assimilating, or rationales for ignoring, propositions from the others. Structural tensions among them also set up barriers to creative interchanges. These define the limits to the art of the possible in integrated global enquiry, and also question the desirability of such projects. Differing interpretations, built respectively on strong and weak versions of the environmentalist account, are nonetheless productive. The first explores the potential for general-theory logics, or new standard accounts, that build economic, social, cultural and other factors on to ecological foundations. In the second, environmental arguments are reintegrated into foundational economic logics, particularly in the context of the processes and consequences of economic globalization. The latter approach points towards theoretically invigorated international political economy (IPE) frameworks that pay greater attention to ecological questions. For the most part I have deliberately avoided singling out the 'global' as a separable level of environmental activity. Globality is either integral to, or a specific application of, both integrated ecological modelling and global political-economy approaches.

2
Scapes, Scops and Sciences

Environmental problems are frequently constructed as interdisciplinary ventures. This strategy is beneficial for global ecological discourse, but it prompts further questions. It is not as widespread, or as deeply applied, as sometimes appears. It tends to be more common in relation to problem- or action-oriented discussions than to theoretical enquiry. Disciplinary bases of research remain crucial for many areas of environmental study. However, these also act as deterrents to more sustained transdisciplinary activities. This chapter explores these problems, and also looks at the differing responses to environmental questions of international relations (IR).

The environmental domain

The topics covered by the terms 'environmental' and 'ecological' are continually shifting and often, I will argue in chapter 4, expanding. Discourse is simultaneously cognitive and affective, and political and social. It urges hunts for problems: for waste disposal from Antarctic tourism or research stations, deterioration of the coral reefs off Okinawa, air quality in Lahore or Jakarta. Environmental assessments vary with economic conditions, localities and the cultural contexts of observers. They are also products of the political calculations of environmental actors. NGOs at different times have used whales and tigers, nuclear-weapons tests, agricultural chemicals and other focal issues to influence public opinion, attract and retain supporters, and structure discourses. Formal transdisciplinary constructs such as environmental science also vary considerably in scope, for example in the extent to which geological and earth-system perspectives are included alongside those of the biological sciences, or approaches from the social sciences are integrated.

The promotion of 'environment words' (Myerson and Rydin, 1996: 37) is central in these processes. Examples are 'warming', 'endangered', 'biodiversity', 'global', 'sustainable', and 'precautionary'. 'Environmental' is itself one such key term. This usage was absent in the nineteenth century, even though many problems, such as those of city sanitation systems and water supplies or the creation of national parks, were in existence and preoccupied polities. Use of the word to integrate and give meaning to diverse activities, demarcate these from other kinds and mobilize public opinion, has thus become a critical differentiating feature of recent decades. The term 'ecological' too is put together in many different ways. It has specific meanings within the science of ecology. Used in political discourses, it tends to connote a preference for setting, or sometimes marginalizing, scientific accounts within economic and social arguments. Its imagery also contains the suggestion, explicitly so in deep ecology, that its insights work at more fundamental levels, and that its prescriptions are more structural, than those spun off by the vocabularies of reformist environmentalism. Unless the context calls for clarification, though, I will use 'ecological' and 'environmental' interchangeably.

Characterizing the environmental as a domain is a useful way of moving between two extremes: regarding this, on the one hand, as constituting the basis for a general or foundational theory of societies and economies, and, on the other, seeing it merely as one of the many sets of issues processed by contemporary societies, or of activities carried out by their members. The term 'domain' is used in a general sense by Layder (1997: 2–4) in sociological theory. He uses it as a means of demarcating broad areas of empirical enquiry, without imposing on them the prejudging implication that one – the individual agent, for example, or larger social structures – is somehow more important as a source of explanations than others. There are similarities with the use of issue-area terminologies in political science. This usage has the advantage of drawing analytical attention to differences in the ways political systems treat different categories of questions. The politics of environmental issues are different from the politics of, say, defence issues. However, issue-area frameworks tend to be oriented more towards political-science or IR questions and methodologies, rather than to the kind of interdisciplinary enquiry required for deeper understanding of environmental problems. Issue-area frameworks also risk seeing environmental politics merely as an instance of single-issue politics, a gloss that overlooks the diversity and internal differentiation of topics on environmentalist agendas, and which reinforces a reluctance to engage with the substantive issues of discourses. Environmental discourse and action, then, form a region of social, cultural and economic space that has its own distinctive features, and complex patterns of connections with other such regions.

This approach leaves deliberately open the question of just how important understandings of the environment are in terms of our capacity to understand economies and cultures. The point is often taken for granted in environmental accounts. Sustainable-development critiques point to ways in which the workings of contemporary economies are out of kilter with appropriate long-term methods of organizing industries and using natural resources. Deep-ecology arguments indicate the serious costs, including perhaps our survival as a species, associated with continued neglect of ecological structures and planetary systems. In neither case, that is, is an environmental viewpoint being put forward that defines limits to its own applicability. In emphasizing the importance of environmental issues, theories of the environment seem to be ineluctably drawn into casting themselves in the role of general theories of societies and economies.

This is not to deny that environmental questions have distinctive or special aspects. Some arise from a long heritage of Western thought and the multiple constructions of nature that have flowed from it. Environmentalist argument is inherently complex. Attempts to identify sustainable relations between human populations and global environments face the daunting challenge of having to integrate knowledge of the realms of the physical, social and personal (Baudot, 1999: 10–13). Other 'special' aspects stem from characteristics of the various phases of environmental politics in recent decades. These features include the scientific and interdisciplinary bases of environmental enquiry, the interpenetration of political with ethical questions, for example over duties to future generations or to non-human species, the transnational character of many issues, the multiple types of actors involved, and a preoccupation with questions of social change and governance. Sjöstedt has argued that analytical and political difficulties arise in international environmental negotiations when actors fail to appreciate the distinctive attributes that separate these from other issues on interstate agendas (1993: 25). It is not clear, however, that environmental questions are unique, or in some fashion different from all others on the agendas of modern polities. A more plausible case is that all issue-sets, however these are broken down, come with their own distinctive mixtures of characteristics.

Nor should we infer that talking about 'environmental' questions requires that there be consensus on environmental issues, or even much commonality in the way these are defined. Although there is an 'environmental orthodoxy' that Easterbrook (1995) and other critics have attacked, in practice contested regions are more common in the environmental domain. Subverting the claims of science has been a hallmark of some species of environmentalism. Defence of the criteria of good science in responses to toxic wastes, declining fish stocks and other issues has been crucial in others. Still others look to alternate epistemologies, such as those of poetry (though in its more traditional un-postmodern senses)(Bate, 2000). A variety of liberal-democratic perspectives on the search for solutions have at times coexisted uneasily in Western states with advocacy of coercive or radical strategies. Environmentalism is associated with the traditional logics of collective action, the voices of members of marginal groups, and the claims of traditional knowledges. It has interacted with other issue-sets to produce a diverse array of hybrids concerned with environmental aspects of gender, peace, North–South inequities and others. Some perspectives on environmentalism betray still the biases and blind spots of the westernism out of which they grew. Others integrated within broader development critiques are more alert to the pitfalls of intercultural transactions. Differing forms of Western environmentalism are sustained by, and support, both the conventional institutions of state, society and economy, and the countermovements of radical change. It is a modern and almost quintessentially Enlightenment project for the improvement of society through the growth of knowledge and sound thinking. However, 'modern' being 'everything we no longer care to be' (Morrison, 1996: 3), environmentalism also takes postmodern turns, or at least postmaterialist ones: it embraces both Keats's angels and those bent on clipping their wings.

The cat-herder's lament

Understanding environmental problems has increasingly been viewed as a project that cannot be handled within the confines of any single discipline (De Groot, 1992: 38; O'Riordan, 1994: 4; Quesnel, 1995; Chechile and Carlisle, 1994). Choucri argues that disciplinary constraints adversely affect study of global environmental problems: 'The time-worn problem is that the assumptions, concepts, theories, and methodologies of the various social science disciplines frequently serve as what amount to "protectionist" barriers that shut them off from one another, with the result that findings do not circulate widely in a common marketplace of ideas' (1993: 2). Yet while much interdisciplinary collaboration has in practice marked approaches to environmental questions, more deep-rooted forms of cooperation that connect the theoretical cores of social-science and other disciplines have been much rarer. Tension remains between environmental enquiry pursued within disciplinary borders and in different forms of transdisciplinary activities. Local dialect and tribal kinship have long been powerful disciplinary bonds. As Aquinas wrote, each science (and he included in the term studies now as diverse as moral philosophy, biology and music) 'has its own appropriate questions, replies, and arguments; and correspondingly its own snares and ignorances'.

New topics, such as environmentalism was in the 1960s, have accordingly had to run the gauntlet of disciplinary gatekeepers. Their authority, and capacity to impose regulatory conditions on newcomers, has varied with such factors as the state of internal dissension in a particular discipline and the degree of fit of the entrant's concerns and capabilities with those already inside. In light of growing threats to the environment, for example, nature reappeared as a central problematic in Western theological discourses in the 1980s. Environmental questions entered philosophy primarily by way of discourses on ethics, and to a lesser extent through subfields such as the philosophy of biology. The process has given much environmental discourse a markedly ethics-oriented flavour. This in turn has fostered contacts with discourses in political philosophy and political theory. Debates on human-rights issues gave environmental questions one of their main entry points into studies of law in the 1980s. Environmental historians took cues from enduring questions such as the changing powers and organization of governments over time and the rise of social movements. In the novels and poetry of Sweden, Australia and other countries, and the musics of composers such as Ives, Messiaen, and Schafer can be found many traces of environmental themes, such as confrontations with or the disappearance of wild nature, that resonate with the imaginative arts' traditional concerns with identities and relationships.

However, of the disciplines with traditional claims in environmental enquiry, including its global aspects, two – ecology (which is a focus of the discussion in chapter 3) and economics, or rather political economy – deserve an early mention. Economics and ecology took their etymological roots from notions of households and their management. In practice, their respective scopes and methodologies were not designed to be similar, or even overlapping, when ecology joined its companion in the 1860s. Each has since diverged into several competing sub-approaches, and each has multiple connections with a diversity of other disciplines relevant to environmental research. Parts of each have established various linkages with parts of the other in the interdisciplinary hybrid forms of environmental economics and ecological economics. I will look first at classical economics as a way into the broader range of concerns of political economy.

Questions that would now be regarded as distinctively environmentalist have a long history in economic thought (Kula, 1998). Among these are problems of pollution control in studies of externalities, and the traditional concerns of economists with resources (broadly defined) and their allocation in conditions of scarcity and competing uses. Many of the areas that economists see as the characteristic advances of the twentieth century – in macroeconomics, econometrics, game theory, and the growth of empirical research resulting from advances in data collection and analytical methods (Baumol, 2000) – also have far-reaching implications for environmental enquiry. Yet in many ways economics remains identified with its nineteenth-century roots in micro-level analysis. The significance for environmental study has been twofold. Firstly, study of the decision-making of individuals and firms in various types of markets is assumed to provide important clues to their responses to environmental problems, as well as their responses to the environmental policies of governments and the filtering through to domestic agents of multilateral environmental agreements. Secondly, the tools required for these analyses – notions of rationality, expectations, utility and choice – have been elevated into a theoretical approach which claims applicability across problems throughout the social sciences, from crime and the organization of legislatures to the manner in which individuals select spouses, careers or religious affiliations, or make ecologically significant choices.

Economics as a discipline has also been subject to internal disputes, and to fragmentation as internal bits fly off at tangents. The process is controlled to some extent by the momentum of the neoliberal project. The agenda rests on markets as the optimal means of allocation in economies, and as mechanisms capable of adjusting promptly to shifts in the expectations and decisions of agents. Associated with this theoretical strategy is an attack, variously qualified in different versions, on states and other institutions that restrict the operation of this logic through such methods as protectionism, state ownership or control of corporations or key sectors, the management of fiscal and monetary policy, subsidies and price controls (Grabel, 2000: 2). Since it also characteristically targets government deficits and debts and inflation as core policy problems, the argument extends to many other areas of the state's relations with economies and societies, from interventionism in general to social programmes in particular. The neoliberal project thus has a direct bearing on many issues of environmental policy. It defines the contexts in which governments formulate and use policy instruments, and their capacities and political will to effect change; it shapes the conditions in which economic agents, particularly corporations, operate; and it both facilitates, and provokes, actions by civil-society actors.

Various strands of institutional economics have disputed some of the assumptions of this broad approach. These give legal and political institutions greater explanatory power, and include theoretical frameworks that start from different bases from the concepts of rational economic agents required in neoclassical economics (Avio, 1999: 511–13). Institutional economic arguments have thus been associated with major critiques of market-oriented approaches to environmental policy in Southern and transitional economies. In Russia, for example, institution-building, in terms of both governments and civil society, has been argued to be a prerequisite to the introduction of effective government programmes and market-based reforms (Sölderholm, 1999: 403–4).

Radical economic traditions take these critiques further. Several features make them particularly relevant to expanding knowledge of the ways societies handle environmental stresses. These include the traditional preoccupations of analysts with institutional factors, particularly the organization of civil society, questions of inequality and power in social relations, the systemic properties of economic systems, the significance of globalizing economies, and with processes of structural change in these systems over time. More specifically, many of the themes in these approaches form a critical link with developments of the 1980s and 1990s in international political economy. As Korany and others have emphasized, IPE has a wide-ranging historical and social scope and transdisciplinary character. Questions of images and identities, of cultures and communication flows, have become more central, and assumptions about the interplay of causalities among structural and other variables more complex. The impact of television programmes from Italy on Albanians in the late 1980s was to create, according to one observer who experienced them, a 'reality where people, and things, and behaviours, and actions are light, colourful, beautiful.' People were 'almost always good-looking, clean, and well-dressed; they all smile and enjoy everything they do, and get extremely happy, even when confronted with a new toothbrush' (Vebhiu, quoted in Nederveen Pieterse, 2000a: 132).

Environmental questions are best approached through study of the interactions between the critical issues of political economy, eclectically defined, and ecology. The focus on globalization in IPE, and the traditional methodological emphasis on the emergent properties of systems in ecology, are key ingredients of this interplay. Each, moreover, takes us into contiguous territory in related fields: into sociology and critical theory, for example, on the one hand, and into other areas of biology and in the earth-system sciences on the other.

We find a variety of adaptive responses to environmental problems in other social sciences. Political science accommodated itself to environmental issues as these grew in significance for Western governments and political parties in the late 1960s and early 1970s, and as environmental groups developed capacities for agenda-formation and influence over governments. The IR subfield, which I will come back to in the next section of this chapter, often took a lead by highlighting the importance of the international developments of that period. Political scientists increasingly argued during the 1970s that issues of the environmental crisis were closely linked to their traditional concerns. Study of environmental politics has thus helped illuminate understanding of enduring topics such as the behaviour of pressure groups, factors influencing voting, bureaucratic politics, and the legislative processes of government. According to Dwivedi, these questions also give political scientists an opportunity to expand the range of their conventional concerns, for example by paying more attention to broader ethical issues (1986). And as Meadowcroft (1997) has argued, political scientists can make distinctive contributions to sustainable-development debates because these require consideration of the institutional factors, such as the capabilities of governments and civil-society groups, that tend to be ignored by many economists.

Means and ends thus became entangled. Political-science concepts and methodologies became *the* means towards better understanding of environmental problems, and analysis of these questions became in turn *a* means towards enriching political-science knowledge. This two-way flow of enquiries has led some critics to argue that the substance of environmental issues has often been lost. Atkinson has suggested that political scientists have largely ignored these, as they prefer instead to look at environmental conflicts as just one of a large number of sets of interest-based politics in Western societies (1991: 21). Related complaints have arisen in analyses resting on epistemological and other bases critical of dominant approaches in political science. Various critical discourses have made use of evidence of ecological deterioration as part of wider arguments about the inadequacies both of state measures in response to economic globalization, and also of the discipline's conventional delineation of environmental questions.

In a sense this is part of a wider phenomenon. Disciplines naturally like to make use of problems that fit nicely into their unfolding narratives. Representational space is sometimes curved. It draws the eye not so much towards environmental or other objects, but back towards the disciplinary ground of the observing figure. And disciplinary communities are not monolithic. Internal disputes are widespread. Sub-communities have their own slants on environmental problems. For example, some versions of ecological feminism have been criticized on the grounds that these fail to promote emancipatory logics (Sargisson, 1996: 211).

Sociology was slower to respond to the environmental debates of the 1960s and 1970s. One delaying factor was probably the fear that ecological arguments might bring in undesirable biological-reductionist or determinist perspectives (Hannigan, 1995: 6–8). Moreover, sociologists have traditionally viewed relations between societies and natural environments as more of a defining feature of social anthropology. Sociology, by contrast, has tended to define itself as the study of modernity, often with a hidden exemptionalist premise that, for those studying such societies, nature was separable and ignorable. As late as 1994 a review of the field concluded that the environment was 'rarely construed in social terms; until recently it featured little in sociological thinking beyond references to the heredity versus environment debate' (Marshall, 1994: 153).

This situation changed during the 1990s (Murphy, 1995; Macnaghten and Urry, 1995; Redclift and Woodgate, 1997; Foster, 1999). The cause of environmentalism was promoted in part by the resurgence of critical sociology. Gismondi, for example, has argued for greater involvement of critical sociology in the epistemic communities surrounding environmental policy-making (1997: 470–1). Environmental questions have become increasingly prominent in several subfields of sociology, and have been examined through a variety of theoretical perspectives. As with political science, sociology has nonetheless tended to make use of environmental questions as a way of examining or revisiting older, or more innovative among recent, issues and concepts. For some analysts this has meant studying environments and environmentalism through the lenses of social class or gender. There was a tendency in the 1980s to view environmentalism primarily as a social movement challenging dominant structures, and as allied, therefore, with other forms of radical politics. Giddens was among the writers who made extensive reference to environmental problems in these terms in his writings of the 1980s. Relatively ignored or downplayed, that is, was consideration of environmentalism as a regulatory arm of the state. The environmental movement was in a sense functionally equivalent to any of the other new social movements. In Habermas, environmental topics recur, though they periodically also disappear in the 'uneasy ambiguity' about nature that characterizes his writings (Vogel, 1997: 176). However, as Dobson has argued, when present they tend to remain secondary to his treatment of human relations and the discourses that take place in these contexts, and are ultimately dependent on them for their resolution (1993: 197–8).

Environmental themes have thus dispersed among many disciplines. The spaces between disciplines have a bearing on these processes. Boundary-maintenance requires constant care of the rules and tools that create difference. Discipline-based criteria provide the means for answering key questions: 'Yes, but is it Political Science?', or: 'What can Sociology contribute to environmental enquiry?' They also guard against encroachments from the novel. As the neoclassical economist tends to respond, 'But it's all in Marshall' (Baumol, 2000: 1–2). These mechanisms influence the divisions of labour that habitually shunt specific questions into particular disciplines (Salter and Hearn, 1996: 136). They also shape perceptions of stratification among disciplines, of prestige rankings based on foundationalist, mathematics-based or other criteria, or, as was historically the case in botany as opposed to, say, physics, of the appropriateness of gender-based divisions among the sciences. Other mental maps construct the observer's own discipline. In biology these increasingly acknowledge that internal disciplinary complexity and growth has surpassed the capacity of its members to comprehend it. Larger cultural factors play a part in these processes, as in the traditional insularities that have at times divided French and German social theory (LaCapra, 1983: 148).

The strong version of the defence of this fragmentation is that diligent care of the paths to knowledge – producing and recognizing good questions and sound methods of enquiry – requires communities of scholars in different fields, and that this value is threatened by interdisciplinary experimentation (Hamilton, 1996: 15, 201–6, 221–2). Hensher paints a withering picture of Ruskin in the nineteenth century 'drawing gleeful links between things, like a lunatic assuring one that, of course, everything is connected' (2000: 35). The complaint is partly that writers lifting concepts or findings from unfamiliar fields can easily misinterpret them or use them out of context. There are ethical problems, too, in harvesting insights from other disciplines that an outsider lacks the capability to reproduce. Boundary maintenance thus implies respect for the rules of interdisciplinary statecraft, including recognition of the status of autonomy in others and observance of the principle of non-intervention. There is also perhaps a suspicion that knowledge games have zero-sum attributes – that the diversion of resources into interdisciplinary research or hybrids saps the energies, the commitment to the elusive concept of rigour, and the cumulation of contributions, of each. A weaker, but more defensible, version is that good interdisciplinary work is likely to be better promoted by protecting the creativity, and even the prickly territoriality, of particular disciplines. Pursuing interdisciplinary options, however, does not so much solve as redefine or defer the nature of the research problem. Choices of approaches still have to be made if the range is expanded to included parts of other fields.

Environmentalist and other advocates of interdisciplinary alternatives can draw on rich veins of evidence of the artificialities of many of the boundaries of the natural and social sciences. Further, the organization of these disciplines in university settings does not reflect the way social and environmental issues are defined and classified outside them. The divergence constrains the potential for discourse among academic, social and other groups on environmental, health, housing and other issues. More thoroughgoing critics have called for fundamental restructurings, as in Foucault's dismissal of the 'so-called sciences', Wallerstein's appeal for reunification and redivision of the social sciences (1999: 21–7), and Taylor's redefinition of the human or hermeneutical sciences. These larger currents have a bearing on global environmental enquiry, but fortunately it is not necessary to wait for the disputes to be resolved before taking this transdisciplinary project further.

In literary criticism, one writer has wryly noted, 'the failure to be interdisciplinary is usually viewed as a serious fault, almost as serious as the failure to theorize' (Levin, 1993: 13). There is a risk, that is, of redefining interdisciplinarity in ways that confirm parochialist biases. Many disciplines are talented producers of integrated perspectives that reflect self-images of their own centrality. A geographer can point to the spatial bases of societies and economies. Political scientists recall their membership of a discipline defined by Aristotle in architectonic terms. Anthropology, philosophy, sociology and economics all in different ways encompass enquiries into topics studied throughout the social sciences. So too does biology, by way of the extension and adaptation of evolutionary concepts.

Environmental enquiry is grounded in particular disciplines, then, but also in two typical forms of interdisciplinary cooperation. The first is the topic-driven strategy. Problems arise within disciplines that require help from outside. The study of plant life, one observer noted in the 1920s, calls for knowledge from physics, chemistry, meteorology, geology and biology (Pearson, 1924: 72). Habermas wrote in the early 1960s, in the context of his attempts to expand understanding of the public sphere, that 'when considered within the boundaries of a particular social scientific discipline, this object disintegrates' (1989 [1962]: xvii). The study of cognition requires cooperation among philosophers, psychologists, neurologists, communications researchers and others. Collaboration on particular problems across boundaries marks many disciplines. A review of sociology has concluded that sociologists 'often have more in common with those in other disciplines working on the same topic than with other sociologists, and it is not always obvious whom to count as a sociologist' (Platt and Hopper, 1997: 283).

Secondly, transborder cooperation may serve to push bits and pieces of different disciplines together in a more sustained fashion. Amalgams and hybrids result. Some take on the attributes of disciplines, like economic history or palaeobotany. Environmental economics has grown out of questions in economics about externalities and natural resource use, and ecological economics from attempts to forge productive links to economic methodologies from ecological concepts. Connections across boundaries are also built by supporters of metaconcepts. Concepts of nature and the elements had this effect in medieval Europe. Notions of system have been pursued as a means of integrating researches in the natural and social sciences, and of information as a way of studying levels of organization from cells to social structures. The integrating role of shared scientific methodologies in such enterprises is usually implicit, but ventures such as the unity-of-science movement in the 1930s occasionally make this explicit. Grand perspectives such as philosophical anthropology in German scholarship in the 1920s, critical theory in its diverse forms, and postmodernism, have suggested alternate integrating or foundational strategies. These become ways of directing enquiry, conceptualizing the observing subject, recognizing fellow-travellers, and connecting ostensibly disparate discourses. Cross-border connections of looser kinds stem from widespread study of particular authors.

Both processes – *ad hoc* problem-solving cooperation and conscious efforts directed towards theoretical synthesis – have influenced environmental enquiry. Both can quickly generate the 'intellectual vertigo' that Ward sees as a characteristic of encounters across disciplinary borders (1996: 2). But there are obstacles. These include what Salter and Hearn call translation problems. 'Each disciplinary (and interdisciplinary) community has a different way of speaking about the topics and the conduct of its research.' Moreover, writers operating 'outside the lines' may not be able to find an audience for their work (1996: 141–6). Even where interdisciplinary cooperation is a professional obligation, as in many areas of the health professions, it is constrained by factors such as vocabulary differences, time limitations, prior disciplinary commitments, and the difficulty of defining common theoretical perspectives (Magill-Evans et al., 2000).

There are normative grounds too for objecting to attempts to bridge the ground between disciplines. Rorty criticizes the accommodative treatment of alternate vocabularies. 'To treat them as pieces of a puzzle is to assume that all vocabularies are dispensable, or reducible to other vocabularies, or capable of being united with all other vocabularies in one grand supervocabulary' (1989: 11). On the other hand, some vocabularies are more mergeable than others. Habermas is less worried about interchanges among disciplines as such – and indeed he draws extensively on political science, philosophy and law as well as sociology – than he is about forced connections among the meta-areas of scientific and technical knowledge and the private and public spheres of individuals' lives (1987: 339–40).

Transdisciplinary connections of various kinds are unavoidable in environmental enquiry. This has an obvious home in the natural sciences, but crucial variables call for investigation in the social sciences. Interdisciplinary scientific study could point to the ways in which toxic chemicals affect wildlife populations or human health, but without research grounded in political economy and sociology we would then still be not much nearer understanding the causes of the human behaviour that sets these processes in motion.

Finally, the location of particular disciplines in larger clusters affects communication flows among observers. The contrasting cultures of the social sciences, natural sciences, and humanities each shape notions of the environmental. In some cases (anthropology, geography and psychology) the fracture-lines run through the middle of disciplines. Natural scientists tend to be respectful of the humanities – *Nature* had a quotation from Wordsworth on its cover when it came out in 1869 – but puzzled by the social sciences. Snow famously pondered, but rejected, the suggestion that they constitute a third culture. There are significant divides within the social sciences. Some areas (in economics and psychology, sociology in its more Comteian forms, and parts of political science) share key epistemological and ontological assumptions with the natural sciences. They are relatively more likely to accept realist arguments about the existence of things, including social things, the relations between words and objects, and the nature of facts and the feasibility of their study by relatively detached observers. Other parts of the social sciences are drawn more into the interpretive and particularist emphases of the humanities, and remain sceptical of the wisdom of seeing human beings as bits of social (or ecological) totalities. Or, in Latour's idiosyncratic definition of the problems of the social sciences, 'objectivity' refers to 'the presence of objects which have been rendered "able" ... to *object* to what is told about them' (2000: 114–15).

These very large-scale knowledge structures have implications for environmental investigation. Assumptions of the orderliness and the recurrent patterns of nature and social life foster compatibilities between many areas of the natural and social sciences on these questions. Environmental economics and ecological economics have built successfully on these epistemological foundations. So has applied research, for example on climate change, that requires the modelling of complex interactions among economic, demographic, social and other variables. But these genres tend to be 'crunchy', to use an *Economist*-word. They are not so hospitable to the more slushily inconclusive modes of much of the humanities.

IR constructs

The responses of IR and IPE to environmental questions have been shaped by their life experiences. Constructions of the history of IR, often used for presentist purposes, are significant features of contemporary debates in the field (Schmidt, 1998: 31). Though now institutionalized as a subfield of political science, IR, like sociology, had an early quasi-autonomous history as a loosely integrated grouping of several disciplines – law, history, economics and geography particularly. Parsons saw it in the early 1950s as a 'synthetic discipline', similar to population studies and other fields 'dealing mainly with contemporary phenomena' (1951: 555). Many of these connections have decayed. Other disciplines have maintained their own international subfields. Sociology later created its own, in part in response to earlier systems theorists, and in part out of the conviction that modernity has inherent globalizing dynamics that require deeper investigation. This growing subfield made productive connections with the research in environmental sociology focused on national societies (Redclift and Benton, 1994). In studying globalization, however, sociologists often 'paid conspicuously little attention to any insights that IR might have to offer on the subject' (Albert, 1999: 239–40).

As part of political science, IR tended at first to screen environmental topics by asking how these could contribute to the understanding of traditional questions of power and influence (Sprinz, 1994; Boardman,

1997; Zürn, 1998). The onus was on those interested in their study and public-policy implications to defend their inclusion. Defenders argued they were 'important' in classical political-science terms – for example, they were significant issues for governments and interest groups. Further, IR is a field with a traditionally, though not currently, dominant paradigm, that of realism. A subset of IR screens historically filtered out many environmental questions which seemed low on issue-hierarchies, unlikely to generate much insight into war or traditional conceptions of security, or which appeared too 'technical'. More recently, as IR and IPE have expanded and diversified, the treatment of environmental topics has been caught up in the mix of crossfire, collaboration and stand-off that characterizes relations among contending theoretical perspectives.

The original impetus for discussion of environmental issues in IR reflected in part the policy-oriented, and largely US-derived, flavour of the field. Commentary on international environmental issues surrounding UN developments in the 1970s nestled comfortably within this tradition. Foreign-policy practitioners took part, as did a variety of interdisciplinary environmental specialists. This pattern has persisted as issues and circumstances, such as the Brundtland Commission's work of the 1980s, and the Rio and Kyoto conferences of the 1990s, have continued to propel environmental issues on to IR agendas. These questions have receded when other grand issues such as the oceans, energy, trade, ethnic conflict or international development have taken the lead. Various factors have nonetheless sustained attention to environmental issues. These include their capacity to shed light on issues on disciplinary agendas such as resource conflicts, transnational NGOs, global governance, North–South relations, and the processes through which many ostensibly domestic-policy issues have become internationalized.

Policy arguments, variously connected with theoretical assumptions about the nature of the international system, have thus been influential in shaping IR/IPE responses to environmental issues. The importance of politics, choice and decision was stressed in critiques of neo-Malthusianism and of the more deterministic interpretations of growth-limits views in the 1970s. Policy preferences for the gradualist reform of international institutions became significant counters to advocates of a neo-Hobbesian or world-government-oriented transformation of the state system. Liberal writers also sided, however, with environmentalist critics of the 1980s in arguing that governments were failing to engage in long-term thinking about environmental problems or to appreciate the challenges these presented, and that the traditional conception and territorial organization of sovereign polities was an inadequate foundation for their resolution. By the 1990s, one of the main lines of cleavage was between writers urging application of sustainable-development principles, and acknowledging that these both required and facilitated a reformed international system (McNeill, 1990), and critics concerned that prevailing interpretations of such ideas represented an extension of Northern and IFI interests, particularly in their dealings with structurally weakened Southern economies (Law, 1997: 176–8).

Three main sets of theoretical perspectives in IR – the respective variants of realism, liberal pluralism, and political economy – have traditionally nourished contending approaches to environmental topics. IPE, though, has in many ways expanded into a rival definition of the field of international enquiry. Further, while theoretical disputes have shaped the contours of IR debates on environmental as on security, human-rights and other questions, the contending parties are not isolated entities. Historically, there has in practice been a significant measure of cross-fertilization and contact among different regions of IR and IPE, and to some extent also between these and outside researchers.

Analyses inspired by the broadly realist tradition of IR reveal complex responses to environmental questions. Realism's characteristic emphases – on power and interest as determining factors in the relations of states, on states themselves as the pivotal actors in international politics, and on conventionally defined high-politics questions of war and security as the foci of enquiry – traditionally raised doubts about the value of investing effort in the study of environmental questions. Much of the stuff of environmental politics lies in the netherworld of culture, education, science and other such things that Morgenthau in the 1940s explicitly rejected as part of the subject matter of international politics. However, definitions of the physical environment in terms of the distribution and scarcity of natural resources have been central to realist analyses as crucial elements in the determination of a nation's power and of the structuring of relations among states. Ecological argument is thus compatible with varying realist approaches. However, realism is not so well able to handle the wider range of environmental and global resource questions, such as climate change, that call for enhanced degrees of trust and cooperation among states and other actors (Laferrière and Stoett, 1999: 104–5).

The history of IR's handling of environmental questions reveals a number of attempts to take such concerns further. Notable among these was the blending of realist and ecological arguments, particularly by way of insights from IR's early connections with geography, in the writings of the Sprouts from the 1950s. Energy and other natural-resource endowments, population change, and the relations between climatic zones and economic systems, were identified as important factors in international politics (Sprout and Sprout, 1962: 378–9, chs 11–13). More obviously 'environmental' themes became increasingly prominent in their writings. These included the consequences of deterioration in natural-resource capital, the asymmetry between one-earth ecological appeals and the territorial organization of humankind, and discussion of the more ambitious cooperative schemes linking states and other actors that were needed to manage global environmental problems (Sprout and Sprout, 1971: 11, 208).

These approaches intersect with others in the classical IR tradition. The concept of interest, especially as redefined in the neo-realist tradition, and the assumption of the centrality of states have been widely used to fit environmental issues into prevailing frameworks. The state thus remained central in its dual role of providing societies and citizens with security and welfare (Holsti, 1995: 67). As issues such as climate change slowly edged up the hierarchies of states' priorities in the 1990s, international environmental politics began to look like familiar exercises in the pursuit by states of their interests (Rowlands, 1995: 151–60). Other actors appeared much less prominent. Earlier, Bull, adding normative concerns to this empirical hunch, had criticized the idea that environmental NGOs had just claim to replace states as representatives of interests (1977: 85–6). The wider security impacts of ecological problems were also questioned, particularly as these were beginning to assume greater prominence in the environmental literature. Deudney (1990), for example, argued against the view that ecological factors were important independent variables leading to armed conflicts.

Divergent tendencies of the classical IR tradition conceded greater potential for governance in a world of states, and allowed varying amounts of space for other actors. States may opt for cooperative strategies, rule-acceptance and delegation of tasks to civil-society actors. Caldwell argued in 1990 (p. 303) that 'an international structure for environmental policy' was already in place. Writers differed in their assessments of how far-reaching the institutional changes needed to be. Options ranged from a continued process of incremental improvement in the UN system and international environmental law, to a radical restructuring of intergovernmental institutions. Nor was there consensus on how centrally organized the regime structures linking multiple environmental 'sub-' areas should be, including arrangements at regional levels. In practice these have ranged from loose schemes deferential to states, to more tightly structured designs that exact compliance and erode the exercise of sovereign powers.

These perspectives subvert the efforts of IR-history typologists to put them into neatly stacked boxes. There is a continuum rather than a dichotomy of viewpoints on civil-society actors. While Young (1995: 38) and others have criticized environmental observers and practitioners for remaining within a largely state-centric discourse, in practice the range of responses to states and inter-actor mixes has been wide. Writers deeply critical of the failures of states to deal effectively with the environmental crisis have argued that governments are insufficiently responsive to ecological argument, excessively attentive to the voices of corporations and international trade and financial institutions, consistently bad at learning and problem-solving, and dangerously neglectful of the long-term interests of individuals and societies. Some critics maintain that NGOs are better equipped to think and act on environmental problems (Wapner, 1996: 152–9). Other perspectives take a more qualified view of NGOs, for example by arguing that, particularly in the South, the need is as much for capacity-building at the state level. Functionalist ideas of governance have historically attempted a resolution of these kinds of problems by proposing that the criterion for actor-selection in discrete policy areas should be capability in relation to specific, largely technically defined, tasks. Often at issue in these arguments has been debate on the nature of environmental problems. These have been portrayed not only as more important than sceptics believed, but also as possessing special qualities that distinguish them from others on international agendas. For example, their holistic and synergistic character, and the artificiality of distinctions between international and domestic matters, means that many ostensibly separate policy problems cannot in practice be disentangled (Paehlke, 1989: 144–5; Sjöstedt, 1993: 25).

Environmentalist critics also began to focus more on the capacity of environmental discourse to generate theoretical insights in the discipline, and of environmental politics to effect a transformation of the international system. Critiques established connections with traditional IR concerns by redefining security in environmental terms. Wars have environmental consequences. Further, ecological deterioration in societies of the South, which may in turn be an indirect product of government actions, can increase the probability and intensity of armed conflicts, for example by exacerbating ethnic tensions resulting from expanded cross-border flows of migrants (Noorduyn and De Groot, 1999).

In IPE as in realist IR theory, environmental themes have been explored in the form of research on politically charged natural-resource questions. Soil erosion in the USA, along with depletion of other natural resources, has been identified as a trend likely to have long-term balance-of-power consequences in international politics (Gill and Law, 1988: 373). Water may become a source of future power politics. More generally, environmental issues cannot be isolated from their political-economy contexts. The decommissioning of nuclear power stations or the imposition of conservation-oriented fuel taxes in OECD countries are clearly connected with factors such as trends in the world price of oil. The promotion of iron-ore mining and other activities accelerates deforestation in Brazil; heavy agrochemical use in Taiwan has ecological consequences for soils, water and health; and industrialization in South Korea has produced serious air-pollution problems (Mittelman and Pasha, 1997: 126–7, 152). Severe costs in terms of human health and ecological collapse have resulted from the chemical industries of Cubatao in Brazil. Ecological deterioration was similarly an increasingly visible factor in Africa's economic predicaments of the 1990s. Debates about the resolution of global environmental problems have thus drawn attention to structural inequalities in the world economy. Cheru has argued that many questions central to the interests of Southern countries, including desertification, toxic-waste dumping, and the environmental implications of debt, have been inadequately treated by multilateral forums such as the 1992 Rio summit (1997: 213–15).

As IR expanded and diversified during the 1990s writers broke new ground, though the pace and directions of change were constrained by what Jörgensen calls the discipline's traditional 'exclusion processes' (2000: 31). Differing strands became linked both with post-Marxist critical analyses of knowledge and power, and also, in the contrasting notion of criticality, with contructivist interpretations that wove their way into IR by way of issues of language, identity and discourse analysis. There was also a revival of more explicit concerns with ethical questions, especially those connected with critical theory. These multiple changes in IR have taken the field as a whole into regions connected with discourses in other disciplines on environmental topics. Yet many of these opportunities have not been taken, in part, perhaps, because of the fading fortunes of environmental topics in mainstream IR outside a fairly restricted range of enquiry on international law and regime developments and environmental security. Similarly, feminist discourses in the discipline have often been centred around critiques of the state and issues related to the agendas of war, economic development and social justice, rather than the interrelated themes of ecological deterioration taken up outside IR in feminist contributions to social and political theory. Developments from Habermas's theory of communicative action, which Risse has argued constitutes an alternative basis for IR theory (2000: 2–7), have a direct bearing on central issues in normative discourses on environmental problems. However, during the 1990s there were few efforts from writers in critical IR theory to make connections with ecological discourses (Laferrière and Stoett, 1999: 162–3).

Part II
Differences

3
Biosphere Grammar

Many scientific disciplines have a stake in enquiry about the biosphere. Ecology has traditionally been central. Competitive pressures, however, have increasingly restricted and redefined its niche, particularly as frameworks built on geological research that link this with chemistry, meteorology and other disciplines have grown into the earth-system sciences. The organizing metaphors of ecological research have survived this transformation, particularly in wider environmentalist debates. As members of a significant, and relatively autonomous, subfield of biology, ecologists have played important roles in structuring environmental discourses. Indeed environmentalist critics outside the discipline, and some inside it, periodically tell ecologists that they should be more active in the public sphere. The Gaia image has been among the constructs providing integrated rationales for such expanded frameworks of global ecological enquiry. This chapter discusses the links between investigations and methodologies from this grouping of natural sciences and broader theoretical approaches to social and economic issues. After reviewing problems in the cross-disciplinary expansion of ecological and earth-system views, I argue that approaches have historically been associated with diverse, and conflicting, perspectives on social and political questions. More specifically, the expansion of global scientific enquiry in the last two decades of the twentieth century has had significant implications for environmental discourses and practices.

Ecology and earth systems

Ecology has traditionally adopted two sets of working principles that have helped define its scope and boundaries. Firstly, ecology focuses on the relations among biological and physical things, and, usually implicitly, makes ontological and epistemological assumptions about the viability of doing this. Darwin's observations on the relations of species set an early context for these developments. Its aim, according to Haeckel in the 1860s, was to examine the relations of organisms and the bases of their 'conditions of existence' (Cittadino, 1990: 87). While plants became a characteristic focus of ecological enquiry in the early twentieth century, ecologists have investigated organisms and their environments in a diversity of settings, from ponds, mossy rocks and discarded beer bottles to volcanoes and the earth itself. However, the meanings of key terms often remained highly contested. For some the pivotal notion of 'ecosystem' or ecological system appeared to refer to something analogous to a complex biological organism, or super-organism, an inappropriate image for more realist scientific critics.

Secondly, ecology has always considered itself a discipline with significant applications. Knowledge of plants is clearly useful to practitioners in forestry and agriculture. The discipline thus formed connections with the arguments that began to appear from the mid-nineteenth century about the consequences of modern societies' exploitation of natural resources (Schoijet, 1999: 516–17). Advocacy became part of the public role of ecologists, particularly in light of the emergence of the environmental movement of the 1960s and 1970s. This requirement led some internal critics to call for greater scientific rigour in ecological research so that policy needs could be met and publics could be educated. For some ecologists, and others including economists and sociologists, this was also a field that had diverse programmatic applications to the organization of societies. Traces of this ambition can be seen in some of the turns taken by environmentalism from the 1960s, for example in deep-ecology and other eco- or biocentric formulations. However, the nature of the guidance ecology could give to citizens and governments has historically been a matter for heated debate. Ecological concepts and propositions have featured as legitimizing devices in ideologies as diverse as proto-fascism, neoliberalism, anarchism and communitarianism.

Though ecological research incorporates enquiry at very different scales, the balance has traditionally been tilted towards empirical research at the micro-level. Ecologists seek out the small and the particular. Pimm, criticizing the tendency, sees them continually going off 'to beautiful, untouched environments [to] study fascinating species' (1991: xi). Given the complexity of natural systems, there is good methodological justification for micro-level case studies. Indeed some have argued that the level should be taken down a few notches further, and that ecologists should develop more interest in very small taxa such as nematodes

(Lawton, 1998). On the other hand, a preoccupation with the micro-level case study can reinforce neglect of emergent properties, or of important systemic questions affecting human societies and economies such as climate change and biodiversity loss. These issues have led to a resurgence of macroecological studies and critical debates on the place of these in the discipline (Brown, 1995; Gaston and Blackburn, 1999: 364). Increasingly, too, many biologists, in conjunction with others from the earth sciences and other disciplines, have argued that study of the two ends of the scale spectrum, from tiny soil or water organisms to big planetary cycles, cannot be separated. Many aggregative processes operate at and link these levels. Microbial processes, for example, appear to act as important controls in interactions between water and rocks, even at temperatures up to 110°C (Furnes and Standigel, 1999).

Ecology has itself become a complex multidisciplinary construct. Conventional definitions of its concerns imply interdisciplinarity. Moore wrote as early as 1919 that ecology is the 'science dealing with the environment. It therefore covers practically the whole field of biology, and is related in one way or another to every science which touches life' (1920: 3). But in a world of expanding knowledges these intentions have also led to a sense of loss of direction. Ecology has been described as 'an elusive, complex, pluralistic, multi-dimensional collection of specialities and points of view' (Cittadino, 1993: 283). The same could be said of many other disciplines, including biology and political science. In ecology, however, the persistence of holistic thinking makes the plight particularly worrisome for its members. A hankering after the supposed theoretical unity of an earlier age, and an urge to recreate something like it out of a fragmented present, has thus recurred in ecological self-examinations since the late 1960s (Pahl-Wostl, 1995: 2, 190; Lawton, 1999: 187). Internal disciplinary change has been complicated by cross-boundary connections. Ecology is an element in large multidisciplinary assemblages, such as the rapidly growing field of conservation biology since the late 1970s (Noss, 1999: 114), and the various earth-system sciences grouped around geology and other disciplines that have advanced understanding of large-scale processes such as the earth's biogeochemical cycles. Methodological diversity, though, has also had its defenders (Weber, 1999). The plea echoes the dismissal in other disciplines of the 'von Neumann syndrome', the unrelenting hunt for some new concept or model with which to unify or revolutionize naturally disparate fields.

Despite, or perhaps because of, this turbulence, ecology has influenced other disciplines and environmental discourse by providing them with both substantive and methodological tools. Firstly, key concepts and propositions have spread widely. They include the applied emphasis on biological diversity that stresses the risks of monospecies cultivation in agriculture and forestry, and the notions of limits and carrying capacity that connect attributes of populations with food and energy resources and other features of their physical and biological environments.

Secondly, the manner of ecological reasoning has shaped public debates and thinking about environmental issues. Ecology is in some ways more important in these wider scientific and social debates as a theoretical approach, or set of approaches, than as a subfield of biology. Holistic assumptions are common to much (though not all) of ecology. They surface in core concepts such as those of ecosystem, community and population. They facilitate the depiction and analysis of systems in terms of trophic levels, and by means of the classical ecological ideas of stability and equilibrium. The doctrine of emergent systemic properties, that the functional whole is more than the sum of its parts, spread rapidly across many fields of biology in the 1930s and early 1940s. Links with universalizing general-systems principles, as in the work of von Bertalanffy, were one upshot. Like other ideas of the biological sciences these later fed into a battery of influences changing the shape of the social sciences. Opponents in the 1930s like Needham, however, thought it 'methodologically impossible' to view the organism and its environment as a whole. Such thinking invited us 'to contemplate the universe in its axiomatic wholeness, analysis of living things being laid aside' (1968 [1936]: 10–11). Its neo-Darwinian and other critics continued to attack holistic thinking as teleological or even quasi-mystical, and to equate ecological research with erroneous assumptions about superorganisms (Ulanowicz, 2000: 114–15). Notions of ecosystems were nonetheless celebrated in the reinvigorated pro-ecosystem consensus of the 1980s as a victory over the backward-looking scientific forces still defending the 'reductionist-mechanistic paradigm' (Patten, 1991: 288). Contested concepts have, however, remained central to wider methodological debates in the discipline, producing in effect a considerable diversity of approaches.

This logic leads to the view that the biosphere – a term first used in the 1870s that ususally refers to the thin film of life and support systems adhering to and interacting with the earth's physical processes – is likewise a complex, interconnected whole. Definitions of ecology increasingly adopted global features in the 1970s and 1980s. In part the movement was an extension of macroecological developments that aimed to promote the study of larger systems. 'Rather than trying to use ever more powerful microscopes to study the fine details of ecological phenomena, macroecology tries to develop more powerful macroscopes that will reveal emergent patterns and processes' (Brown, 1995: 11). Moving beyond this to a planetary scale, however, posed major problems for the traditionally small-scale orientation of ecological research. General ecological principles helped with this translation. Population-resource dynamics can in principle be modelled and quantified whether the focus is seabirds on a cliff or humans on earth. However, while the relatively simpler language of ecosystems used for micro-level analysis is useful at this grander level as an orienting metaphor, it tends to fail when confronted with the complexity of earth systems.

The globalizing of ecology in the 1970s and 1980s thus reopened the issue of interdisciplinarity. Other disciplines were obviously crucial to biosphere enquiry. The atmospheric sciences became indispensable for work on climate-change issues as these came to the front of scientific and public agendas in the 1980s, as did interdisciplinary Antarctic science for investigating ozone-layer depletion. There has been growing interest by geologists in acquiring environmental data from the recent geological past with a view to contributing to sustainable development (Mathig and de Mulder, 1998: 37–8). The collaboration of several disciplines is required for research into the global carbon, nitrogen and hydrological cycles and other earth-system processes. Transborder, regional and global-level research also highlighted the need for investigation of economic, demographic and other variables, as in studies of the transport of pollutants. High concentrations of ozone are part of the process of formation of photochemical smog. Its study and politics were traditionally focused on specific urban centres such as Los Angeles, but research has increasingly located problems in wider geographical contexts such as eastern North America, or large areas of Europe, because of growing understanding of the mechanisms of the transport and accumulation of ozone (*Atmospheric Environment*, 2000: 1857). The modelling of ocean-atmosphere processes in General Circulation Models (GCMs) from the late 1960s has been a significant contribution to such investigations.

In sum, these developments brought with them the prospect of the interdisciplinary global project achieving three objectives: firstly, quantifying the transformative effects of humans on their environment; secondly, assessing limits and functions, or the degree to which the pursuit of economic and social goals is respectively constrained and facilitated by the operation of ecological and earth-systems laws; and, thirdly, judging the variance of environmental problems that can be explained by reference to human actions as opposed to naturally occurring phenomena. The last was a particularly contentious issue in debates on climate-change issues from the mid-1980s. At opposite ends of a scientific and policy spectrum were scientists pressing the anthropogenic case and urging restraint in the emission of greenhouse gases, and critics arguing not only that there were methodological and data flaws in this argument, but pointing too to natural variability in climate conditions over geological time.

If the biosphere can be considered a system, or set of interconnecting systems, perhaps the earth itself is characterized by the kinds of self-regulatory processes found in systems at other levels. The supposition, fanciful or metaphysical to some of its critics, is central to the Gaia image. This emphasizes historical interplay at both grand and minute levels between the earth's biological and physical processes. In part it represented an attempt by Lovelock to respond to what he saw as puzzles in the earth's geological and biological history. Whereas the O_2 content of the atmosphere has resulted historically from living organisms, for example, its maintenance at around 21 per cent should be seen as surprising. Similarly, while there have been significant fluctuations in geological history in radiation from the sun, the earth's surface temperature has remained relatively stable. Phenomena such as these prompted the idea that the biosphere was an active force in regulating physical and chemical processes (Schlesinger, 1997: 13–14). According to Margulis and Lovelock, the term was chosen to emphasize the multi-disciplinary character of planetary ecology. The term ecology was thought inappropriate because of its history as a subfield of biology, and also because of confusions in the use of the term outside that discipline (1989: 2–4). Important features of Gaia imagery include stress on the role of biota in regulating geochemical processes, and more generally promote enquiry into the mechanisms by which macrosystem stability is maintained (Margulis and Olendzenski, 1992).

The metaphor inevitably provoked complaints about teleological, and tautological, reasoning similar to those that had greeted ecosystem thinking half a century earlier, especially perhaps as it soon attracted widespread interest among environmentalists. Critics argued that the Gaia image lacked the capability to produce empirically testable propositions, a point Lovelock strenuously denied, and suggested that the idea of earth systems maintaining stability through a repertoire of responses to perturbations hinted at some hidden notion of consciousness or purpose at work. A possible corollary was the ethically unpalatable view that the survival or well-being of one species, humans, or of a significant number of them, might be the cost of these system-maintenance operations. A related pessimistic implication was that human actions designed to mitigate anthropogenic ecological stress of the kind that was causing global warming might be so shot through with unforeseen consequences resulting from systemic processes that the rationality of environmental policy and remedial actions became questionable. Lovelock, however, rejected this kind of inference, arguing that whether the consequence is prosperity or extinction, or something in between, depends on the nature of the actions taken (quoted in Ahuja, 2000).

Towards social ecology

To what extent can ecology and the interrelated earth sciences explain the workings of societies and economies, or contribute to the making of good environmental decisions by civil-society groups and governments? The strong version of the answers to these questions insists that these disciplines (or particular mixes of them) are foundational. A greater sense of disciplinary relativity is found in weaker defences, such as those calling for collaborative or hybrid arrangements among the natural and social sciences.

As we saw earlier, ecology has always defined itself in part as an applied discipline. It sees itself as having useful things to say about problems such as the use of pesticides and other toxic chemicals, the management of forestry and fisheries, agriculture, wildlife conservation, climate change and energy consumption. The cognate field of conservation biology has grown in parallel with ecology as a response to alarm about problems of endangered species and habitat modification. As Polis has put it, 'one crucial function of ecology is to provide an unbiased, scientific basis on which political and social decisions can be made about how best to treat our natural environment' (1998: 745). Ecologists, and biologists more generally, remain divided, however, on critical questions concerning the relations between scientific research on the one hand, and participation in the solving of society's problems on the other (Norton, 1998). The calls of environmentalist outsiders for remedies to complex problems of natural resource management or biodiversity often seem strident, or sharply at odds with the precepts of scientific caution. As a result, as one internal critic has put it, 'the political role of ecologists has fallen increasingly to non-scientists' (Peters, 1991: 12).

Ecological concepts have in this process of transmission become enmeshed in diverse ideological constructs. As Bramwell wrote in the 1980s, ecology 'is now a political category, like socialism or conservatism'

(1989: 39). The adaptation has involved some neglect of the complexities of natural systems and the dynamics of ecological change (Vayda and Walters, 1999). One historical translation took the holistic premises on which a large part of the ecological enterprise is founded, and used these to displace individuals and assert the prior claims of the collective social whole. Haeckel's own approaches in the nineteenth century have been criticized for helping to foster the later growth of right-wing authoritarianism in continental Europe. Ecological insights reinforced, or gave a spurious scientific legitimacy to, background influences such as the cultivation of the forest ideal found in nineteenth-century German writings, and the 1930s idea of the 'harmonic landscape' that situated individuals in totalitarian conceptions of nature (Gasman, 1998: 3–4, 380; Weiner, 1992: 387–8). Ideas of the holistic character of nature, and the significance of this world for social organization, became important ideological elements in school curricula in Germany at that time. The later discovery by environmentalists of Heidegger's writings on nature and science prompted similar controversy. Other routes out of ecology into society led to social Darwinism, for example in the context of capitalist expansion in the United States in the late nineteenth century. More recently, different interpretations of ecological concepts and propositions in environmental discourses have identified highly decentralized polities, or networks of small interconnected communities, as the types of social organization most sympathetic to ecological visions. There has also been a resurgence of redefined 'harmony' notions. These have led variously to critiques of capitalism, idealized images of aboriginal polities, a sacralization of nature, or promotion of the traditional knowledges of communities in countries of the South. Ecology, it would appear, is an almost infinitely pliable tool in the hands of social engineers.

Particularly influential in the last quarter of the twentieth century was debate inspired or provoked by the limits-to-growth publications of the 1970s. These criticized the assumptions of technological optimism and unlimited carrying capacity held to underlie prevailing economic-growth models and practices. Regional catastrophes in South and Southeast Asia were predicted as the prelude to an early twenty-first-century global crisis. In a way more important than the specifics of the FM (Forrester/Meadows) global-modelling exercises, however, was the widespread debate on left and right, and in North and South, that greeted the publications (Schoijet, 1999: 520–3).

Looser interpretations of ecological concepts, moulded by environmentalist accounts, have found niches in the conventional political discourses of Western states. Liberalism, conservatism and social democracy, and their respective variants, each have strong ideological as well as pragmatic-electoral grounds for both accommodating and resisting environmental arguments. Ecological ideas have also historically had a mixed reception in the larger critical discourses of modernity, whether Marxist, feminist or post-structuralist, though they have found a more unequivocal defender in social ecocentrism. Some analyses have equated the protestations of many environmentalists with those of the bygone utopian socialists criticized by Marx. They are held to fit into the category of aesthetic or romantic regrets about the consequences of capitalism, that is, or of shallow critiques that hide a misguided belief in its reformability (Martinez-Alier, 1987: 218–19). The belief that the present age is one of environmental crisis is not incompatible with this tradition. Indeed, depiction of the ecological condition of late modernity in the language of crisis is a perspective that has drawn productively on conventional class-based methodologies for the analysis of the hierarchies of contradictions in capitalist economies. At stake is not so much a recognition of the facts of ecological deterioration as an understanding of their aetiology, and, consequently, of the strategies for remedying them. Ecological-Marxist approaches maintain that much green discourse misses the point by failing to grasp late capitalism as the root cause of the global ecological crisis (Kovel, 1995). This tradition thus up-ends conventional ecological argument by making both it, and the conditions it describes, contingent on the economic forces and productive relations of industrialized economies.

Ecology's claims as a social-theory producer, as well as its capacity to fulfil its 'canary-in-a-mineshaft' role, are thus weakened in several ways. Some scientific critics of ecological research have criticized what they see as lack of rigour and a tendency to weak inference. An additional source of complaint, especially among social scientists, has been insufficient attention to humans and their interactions with nature. This criticism was a formative influence in the creation of the hybrid discipline of human ecology. It echoes an old complaint of biology's comparative neglect of human activities. Subsistence hunters, for example, exert significant impacts on their natural environments and the interrelations of wild species, but these effects tend to be marginalized in biological research (Jorgenson and Redford, 1993: 368). Effluents from factories and farms affect river, lake and estuarine ecosystems. Many of the interesting analytical questions here relate not to these multiple effects, but to their economic and social causes. Biology traditionally has a limited capacity to handle such questions, or at times even limited interest in them.

The critique from social theory and the social sciences, then, is that when applied on a macrosocietal scale ecological reasoning tends to fall back either on determinist or reductionist arguments that minimize complexity and choice, or on taken-for-granted assumptions about human behaviour. Individual actions and the structures and processes of societies, as one sociologist has complained, come to be seen as little more than the side-effects of the workings of 'majestic ecological laws' (Wolfe, 1993: 83). These flaws taint several of the 'deeper' varieties of thought in social ecology, Lucardie argues, as these imply the 'subordination of mankind as a whole to the Community of Life' (1993: 22). Neglect of choice is particularly disconcerting to rational-choice theorists such as Elster who, especially in his earlier writings, argued that biological reasoning is inadequate for the task of studying the strategic interactions that are the crucial aspect of human behaviour. Ecologists have applied game theory to questions such as the contests of spiders or birds for food, territory or mates, and to studies of cooperative behaviour. For example, the more intense the begging behaviour by nestlings the greater the quantity of food received, but this requires relatively high investment in terms of oxygen consumption. Situations such as these seem ready-made for game-theoretic applications, but some big assumptions are required before they can be viewed as analogous with strategic decision-making in humans.

There is environmental unease too with the excessive caution of professional ecologists in relation to socially and economically important questions. While biologists, palaeobotanists and others study the geological history of species extinctions as part of 'natural' processes, ecologists have been criticized for failing to grasp the enormity of contemporary problems. Pimm, for example, has referred in this context to the 'expected catastrophic extinction of species [that] will alter the planet's biological diversity so profoundly that, at known rates of speciation, it will take millions of years to recover it' (1991: xi).

Ecology provides analyses of environmental problems and contributes to their solution, then, but its lessons are typically filtered through diverse sets of cognitive screens. As social commentary, ecology often seems to come only in a variety of hyphenated forms as key ideas are selected, absorbed, ignored or rewritten by liberal pluralists, authoritarian planners and others. Images of the relations between science and public policy are also part of the applied personality of ecology. These support key beliefs, for example the view that the creation of efficient communication channels from scientists to publics and politicians – conveyances de-clogged of 'politics' and misunderstandings of the nature of science – is a sufficient as well as a necessary condition for sound public policy.

Global processes

Ecology and the earth-system sciences contain distinctive, and often policy-relevant, views of the factors at work in critical global structures and processes. In looking at these it is useful to relax for a moment the characteristic social-science assumptions that stress the importance of economic and social variables, for example in the use and allocation of natural resources. This temporary suspension gives us a clearer picture of the arguments for the potentially foundational character of ecological analysis, as well as of the limitations of these kinds of theoretical approaches.

In such approaches, ecological and earth-system constructs are used to generate frameworks in which the factors constraining and facilitating the operations of economies and societies arise out of processes of the physical and biological environment. Planetary image-making reflects rapid growth in global scientific research, particularly from the International Geophysical Year (IGY) of 1957–8, and continuing through large-scale multi-country collaborative efforts such as UNESCO's Man and the Biosphere (MAB) programme and the International Geosphere-Biosphere Programme (IGBP). While these activities have represented a tiny fraction of overall scientific research on related problems, they have nonetheless been significant in generating wider public as well as scientific interest in global issues. Central to this research have been assumptions about the utility and feasibility of studying the earth as a system, or rather as a complex set of interacting systems, with 'its interior, its oceans and its atmosphere [forming] an interwoven, continually changing fabric that links inorganic matter and living processes' (Drury, 1999: 93).

A primary disciplinary impetus has accordingly been from geology and the study of dynamic earth processes. Much of the subject matter of geology falls outside the conventional environmental-discourse domain. Longer-term projections into the future are also less obviously relevant. We could speculate that monitoring Australia's slow progress into Asia, or the future rift-based break-up of east Africa, will eventually lead to understanding of geological events of considerable ecological, economic and social significance; but the pace of the movement of the earth's plates, combined with the absence of human causal factors in these processes or capacity for their management, means they usually cannot be constructed as 'environmental' issues. Other geological topics, though, particularly volcanoes, earthquakes and study of glaciation and interglacial periods, are more directly relevant to economic and social issues. Volcanic eruptions, like those of Mount Iso in Japan in 2000 and Mount St Helens in 1980, have environmental consequences, and are also handled differently by governments and civil-society actors from disasters attributable to human activities. Future larger ones have the potential for serious climatic effects with extended biological consequences. Historical study of past periods of earth warming, and of the geological history of changing balances of gases in the atmosphere and their interaction with life processes, illuminates contemporary climate-change debates. Further, shorter-term geological or geoenvironmental changes – over the last few thousand years, or even the last few centuries – have increasingly been recognized as important for understanding environmental problems associated with the changing composition and dynamics of the earth's surface regions.

Developments in geology have thus become increasingly interconnected with questions from chemistry, biology, and other fields regrouped as components of the larger fabric of the earth-system sciences. The process has also reinforced disciplinary fragmentation and subfield formation inside geology (Ernst, 2000; Schellnhuber and Wenzel, 1998). The upshot, according to Schellnhuber, is the making of a second Copernican revolution. 'This new revolution will be in a way a reversal of the first: it will enable us to look back on our planet to perceive one single, complex, dissipative, dynamic entity, far from thermodynamic equilibrium – the "Earth system".' An impetus to this revolution is 'the insight that the ecosphere's operation may be being transformed qualitatively by human interference' (1999: C20).

Several sets of topics are relevant to discussion of linkages with economic and social developments and for comparative assessment of the theoretical significance of earth-system perspectives. I will look briefly at the earth's grand cycles, ecological functions underpinning the workings of economies, landscape and biological diversity, and comparative and systemic processes.

Earth cycles

Linking the atmosphere, biosphere, lithosphere and other large-scale structures are processes that cycle elements. These work through various combinations of anthropogenic and natural forces, and have a variety of beneficial and hazardous consequences for humans. The most prominent, and currently most politicized, of the earth's cycles is the carbon cycle.

This is one of a number of cycles that interact in multiple ways with economies and societies. Scientific concern about the effects of increased CO_2 concentrations in the atmosphere resulting from expanding industrialization began to be voiced as early as the 1890s (Levis et al.,

2000: 1313). In climate-change debates a key focus has been on the impact of fossil-fuel combustion, particularly from Northern economies (Hansen et al., 2000). Carbon-cycle politics has focused on the feasibility of reducing global levels of production of CO and CO_2, and those of other greenhouse gases (GHGs), by emissions-trading and other methods. Policy debates have extended into other regions of the cycle, for example into discussion of instruments for enhancing carbon sinks by forestation. Such options were politically attractive to some Northern states in 1990s climate-change negotiations as a way of avoiding costly energy-consumption and other adjustments to GHG emissions regimes. However, there were objections from the South, and from environmental groups, on the grounds that Northern economies would in these scenarios just continue with past fossil-fuel-use practices. Critics also objected that reforestation schemes, for example in Thailand, have had uncertain or damaging consequences for agriculture and raise difficult land-ownership questions (Grubb, 1999: 98–9). There is also scientific uncertainty on key questions such as whether northern temperate forests in Scandinavia, Russia and Canada act more as a carbon sink or a source of emissions. A significant issue inseparable from political discourse on carbon-cycle issues has thus been the role of factors other than anthropogenic GHGs in producing change effects. Wetlands, for example, produce about 20 per cent of global methane, also a GHG (Saarnio et al., 2000).

While the connections between these processes and climate change were established in the scientific and policy consensus of the 1990s that led to international negotiations, interpretations of the nature and robustness of the linkages have continued to be surrounded by uncertainty (Pahl-Wostl, 1995: 196–208). Singer, one of the strongest critics of the anthropogenic argument, argues among other things that the computer models do not work; the complexity and variety of human influences on climate is downplayed; the good as well as the bad consequences of climate change processes are ignored; policy prescriptions envisage burdens of adjustment in terms of energy-use patterns in the North and standards of living in the South that are unacceptably high; and that alternative methods of both adaptation and mitigation exist (1998: 287–9). There are uncertainties, too, over the geographical spread of causes and effects. With some important exceptions, most countries in the South are relatively minor producers of GHGs from fossil-fuel combustion and other sources. Yet these, for example Uganda (Bwango et al., 2000: 150), are nonetheless affected by global climate change and other effects resulting from processes in Northern states and by regime developments.

The nature and effects of change are uncertain. Speculation on possible consequences plays an important part in climate-change policy debates on adaptation and mitigation. For example, increased CO_2 in the global atmosphere presumably leads to patterns of increased plant growth; and altered vegetation patterns in temperate biomes would have widespread consequences for agriculture, forestry and other land-use activities. Research on past glacial–interglacial cycles has also been an important means of clarifying the nature of climate-change processes in the present. For example, one of the warm interglacial intervals of the late Pleistocene period appears to have been associated with higher sea-surface temperatures and coral-reef expansion (*Global and Planetary Change*, 2000: 1–3). The 'contemporary' period in this context is study of climatic variations and vegetation and other responses over the last 21,000 years, or since the last glacial maximum (Kershaw and Whitlock, 2000: 1).

The beginnings of a similar kind of politicizing process are evident in relation to other cycles, particularly nitrogen. As the most abundant element in the atmosphere nitrogen is critical for the survival of animal and plant species. It is affected by diverse economic activities, including agriculture and fossil-fuel use. The cycle covers a variety of processes, for example through plants taking up ammonium and nitrate. It has been estimated that human activities have approximately doubled the rate of nitrogen input into terrestrial nitrogen cycles. Possible cumulative effects include contributions to loss of nutrients in soils, acidification of lakes, loss of plant biodiversity, and declines in coastal fisheries because of changes in estuarine and nearshore ecosystems (Vitousek et al., 1997). There are also interactions with other processes to consider, for example the impact of increased CO_2 on the nitrogen cycle through effects on soils resulting from enhanced photosynthesis (Hungate, 1999: 280–1).

In relation to some of the other large cycles emphasis has been placed less on mitigation and more on adaptation by human societies and economies, or a wait-and-see approach. The latter is associated with study of complex ocean–atmosphere interactions in the hydrological cycle and other processes. Possible changes in the Labrador current affecting the Gulf Stream resulting from increased Arctic temperatures and consequent melting of polar ice have been predicted in some global-change models. Adaptation to El Niño events by way of increased scientific understanding was a feature of the 1980s and 1990s. There were eight severe events in the last hundred years, and others have now been identified for the past several thousand years (*Earth System Monitor*, 1999: 3). By the time of the 1997–9 El Niño/La Niña events, provisional forecasts became feasible of the likely weather consequences for North and Central America. Research indicated the greater likelihood in these regions of extreme conditions in winters immediately before El Niños, and so furnished estimates of the possible economic implications, for example in terms of altered patterns of energy consumption. Even in the case of the 1986–7 El Niño scientists were able to issue warnings to farmers in Peru and advise them, for example, to prepare for the event by planting moisture-tolerant crops (Schoonmaker, 1997). But the costs of such events have still been significant in Peru, Ecuador, Chile and Argentina in terms of destruction of homes, the spread of disease and effects on wildlife. And effects are experienced widely. In Uganda, rainfall can be delayed by up to two months in El Niño years and rainfall amounts increased in some localities by over 200 per cent, with significant results for crops. Other consequences in East Africa have included cholera outbreaks in 1997–8 resulting from floods and landslides related to El Niño, and damage to transportation systems (Bwango et al., 2000: 149, 151). Ocean–atmosphere dynamics are complex. Patterns of change, including statistically unusual or extreme weather conditions, appear to result throughout the El Niño-Southern Oscillation (ENSO) cycle. Studies have pointed to relations between precipitation events on the Pacific northwest coast of the USA and ocean–atmosphere convection processes around Indonesia and the western tropical Pacific (Higgins et al., 2000: 813).

Ecological functions

Environmental discourse has traditionally centred on such topics as the unintended costs of actions, for example in relation to the future availability of resources. The limits-to-growth arguments of the 1970s drew on the historical record of natural resource-use trends and pointed towards scarcity crises in the early-to-middle twenty-first century. Population-growth debates remain linked to uncertainties about the earth's carrying capacity, especially in light of changes in technology and very unevenly distributed patterns of production and consumption (Cohen, 1996).

A more dynamic, ecologically grounded way of viewing these concerns is to look at the functions of global ecological systems. Services to economies provided by these are varied, and moreover are normally not priced in markets (Daly and Townsend, 1992; Daily, 1997). One rough estimate has put the global value of these ecosystem-economy processes in the range of US$ 16–54 trillion a year, with an average of US$ 33 trillion (Costanza et al., 1997: 257–9). The kinds of services identified are the regulation of gases in the atmosphere, water supply, nutrient cycling, pollination, the provision of raw materials, and recreation. The economic and social significance of these resources and processes has been emphasized in critiques of the lack of such key measures in traditional GDP and national accounting (O'Hara, 1999: 84). The development of national environmental indicators has been one response to this gap in Canada and other countries. Indicators have been designed for tracking key phenomena over time in areas such as ecological life-support systems, human health and well-being associated with air-quality and other phenomena, and the sustainability of natural resources (Neimanis and Kerr, 1996: 373).

Landscape and biological diversity

Changing atmospheric regimes interact with biological processes in complex climate-vegetation systems. Among other things, these result in threats to wildlife species and habitats globally. More severe threats occur in high-density hot spots such as parts of Indonesia, Brazil's eastern forests, and Madagascar. The dynamics of such processes can be studied through use of a variety of land-surface models that link biosphere and atmosphere processes. These have paid increasing attention to climate change and the effects on and further consequences of vegetation changes (Verseghy, 2000: 2–3). For example, climate change resulting from increased GHG emissions has consequences for vegetation, and changes in vegetation cover in turn have effects on local and regional climate systems through interaction with the hydrological cycle, and consequences for radiation from changes in land surfaces. Expectations of increasing CO_2 concentrations in the atmosphere thus lead generally to predictions of the increased geographical extent of vegetation and changes in canopy density (Levis et al., 2000: 1321–2).

There are wider geomorphological consequences of these interactions. Tropical rainforests respond to rainfall, and responses shape patterns of soil erosion, landslides, and so on. Effects are mediated through particularly important biomes in geographical regions such as Amazonia and other parts of the tropics. These act to 'clean' the atmosphere by removing trace gases, a function increasingly threatened since the 1980s by accumulating concentrations of atmospheric methane (Crutzen, 1987: 126–7). Given the global character of anticipated climate-change factors, moreover, different effects potentially extend across all countries. Australia has a high economic dependence on agriculture, temperate forests and other resources. It is thus likely to be affected by a diversity of climate-change-related factors such as changing insect populations and fires. The responses of species to climate change vary, however, certain forest species being less affected than others (Williams, 2000: 68).

Changes in sea-surface temperature distributions regionally and globally have been identified as among the factors that affect atmospheric circulations in West Africa, and that hence contribute to drought conditions in the Sahel region (Wang and Eltahir, 2000: 795).

These processes have a bearing on the increasingly global enquiry into declines of freshwater fish, amphibians and coral reefs. Threats to freshwater-fish populations in Canada, for example, are increasingly linked in the early 2000s to global warming. While a variety of local factors determines processes of change, the geographical extent of decline in such cases indicates influences on a larger scale. Severe declines in local amphibian populations in Australia, South and Central America, and high-altitude areas of the USA have been observed since the 1970s. A number of causal factors have been suggested, including local climate changes, chemical pollution of waterways, acid precipitation, diseases, and increases in ultraviolet (UV-B) radiation. Recent research has indicated that the pattern appears to be global in character, with particularly significant declines during the decade from the late 1950s (Houlahan et al., 2000). Coral reefs worldwide are declining in relation to a variety of global environmental stresses and other factors. These include increased CO_2 concentration in the atmosphere, growing human populations and the impacts of economic activities, changing carbon- and nitrogen-cycle flux rates, and the spread of chemicals, for example from pesticides (Thomas and Dodge, 1999: 12).

Comparative and systemic processes

Mechanisms linking the earth with external factors and processes are environmentally significant. The proportion of the sun's radiation that is not reflected back into space or absorbed by atmospheric components hits land and water surfaces, and part is re-radiated upwards to warm the atmosphere in the so-called greenhouse effect. Related processes shape vegetation through photosynthesis. Sun-spot cycles, part of the tale of nineteenth-century economic thought, may have a possible role in climate-change processes; and the moon too may influence temperature cycles on earth through its governing role in relation to tides. Other solar-system processes generate issues relevant to discourses on the physical environment of economies and societies. These include effects mediated through the ionosphere (Lyon, 2000). There has been growing scientific interest in threats from the nearly 40,000 tonnes of comet and asteroid debris that hit the earth each year (Brownlee, 1998). The collision of comet Shoemaker-Levy 9 with Jupiter in 1994 reinforced media and scientific attention to the issue, as did the near-misses of Earth by asteroids

1999BJ8 in early 1999 and 2000QW7 the following year, and historical research on the Tunguska explosion over Siberia in 1908. Estimates vary of the probability, avoidability and extent of damage from impacts. The hazard has been described as 'the most serious problem humanity faces because it can destroy society globally at once' (Gehrels, 1999: 1160). The research connects with global-change enquiry in various ways, for example through investigation of catastrophic species-extinction events, like that at the Cretaceous/Tertiary (K/T) boundary of 65 million years ago, and comparative extrapolations from knowledge of Venus or from early life-history similarities with Mars to understanding of climate-change processes on Earth. Study of stellar life cycles, of course, suggests that in the (extremely) long term, talk of protecting the earth's future is bound to be fruitless.

Geosciences and global ecology

Although ecology has had a traditional affinity with many of the topics of environmental discourse, no single science treats these in comprehensive fashion. Integrated groupings, however, such as the frameworks of topics conventionally incorporated within definitions of environmental science, present an additional range of problems. These are characteristic of any set of links between disciplines with differing scopes and methodologies. Use of the geophysiological metaphor in Gaia imagery links complex biological, atmospheric and physical and chemical processes, and represents an ambitious attempt to secure a more deeply integrated synthesis. However, this in turn reopens many of the older methodological issues raised traditionally by ecological research, by its holistic approaches more generally or, more specifically, by problems arising from macroecological searches for large-scale emergent properties of systems.

These groupings of sciences nonetheless constitute a significant route into the study of global change by extrapolating to that level concepts and propositions such as those on population–environment relations in debates about the earth's, or specific regions', carrying capacity. Underlying notions of systemic interconnectedness, combined with inbuilt scientific principles of caution, also form a basis for environmentalist defence of the precautionary principle in policy approaches to toxic chemicals, land use, changing biotechnologies and other questions. Scientific enquiry into biogeochemical and other cycles likewise reinforces global-commons arguments.

Some integral limits, however, constrain the application of such insights to more general accounts of economic and social change. Many of the traditional blind spots of the sciences persist, especially the relative neglect of approaches and methodologies from the social sciences and humanities. Without these the implicit causal chains of much scientific research on environmental topics have significant gaps. Explanation, as opposed to description, of the human behaviours that produce effects on environments is missing. These factors are not normally denied, as in forms of determinist or reductionist arguments that remove the factor of choice. But gaps tend to be filled with taken-for-granted assumptions about why people or companies pollute or destroy wildlife habitats, or about the factors that govern people's or governments' responses to environmental or scientific communications. Crucial among these beliefs, too, is the implicit proposition, not always well founded, that analysis of the mechanisms of ecological change will, through a rational process of social deliberation and government action, produce appropriate problem-solving responses by societies.

Ecology, whether alone or in the context of the related earth-system sciences, lacks clear and uncontested relations with specific social, economic or political ideas. At different times and in different cultures, its principles have been used to support disparate arguments about limits or bounty in approaches to the use of natural resources, conflict or cooperation in accounts of the implications of ecological factors or models for understandings of social processes inside and across the borders of countries, and cosmopolitanism or nationalism in the territorial organization of earth's societies. There is a wide range of rival conceptions of the optimal organization of polities. Outside these disciplines, the value of scientific research is also disputed in some cultural critiques. It is portrayed as associated with externalizing and objectifying ontologies that reinforce the temptations, and help create the technological possibilities, of an eventually self-defeating strategy of control of nature. Even within environmental discourse there is often a reluctance to engage with scientific research findings on their own terms. Many of these issues are raised by enquiry into alternate representations of global ecological processes, which is the theme of the next chapter.

4
Ecological Representations

People act in ways that sometimes, perhaps most of the time, have direct or indirect consequences for their natural environments. They may do this out of habit or as a result of careful deliberation. But they also write or talk about their actions and their environments, describe or experience them in novels, films and music, and feel optimism, serenity or dread when contemplating nature and its resources. In doing so, they may use 'environmental' language and styles to describe what they are doing and the consequences of their actions. Or, without resorting to such devices, they may nonetheless act in ways that someone else, looking in from the outside, thinks has ecological significance. There is a wide spectrum of theoretical responses to the question of whether, and how, all this matters. I first look at these in the context of understandings of discourses generally, and then focus on the multiple settings and themes of environmentalism and global enquiry.

Narrative vs./as reality

Responses to these kinds of questions vary from the realist position, with its assumption of an objective and knowable world external to the viewer, to constructionist perspectives that put stress instead on the manner in which observing subjects create their social and physical worlds. The narrative itself then becomes the functional equivalent of the realist's out-there world. Talk, or for others writing, becomes the pertinent practice and the reality. As against the modernist view of the task of understanding the out-there reality, the focus switches to language. 'One doesn't learn abstract concepts and then figure out how to apply them; to use a form of language is itself to engage in a practice, and this usage can have widespread impact' (Gergen, 1999: 167).

In between, or in addition, there are a variety of apparently middle-ground views. These include attempts to build on the importance for social life of actors' perceptions while deflating what one critic has called the 'extreme constructionist position' (Hannigan, 1995: 3). Sabini and Schulkin argue that there is no fundamental conflict between exploring social constructions on the one hand, and biological realities on the other; both, that is, are real (1994: 212–13). This approach has been associated with traditional uses of image theory in IR. By objectifying perceptions, however, and treating these as data and hence as appropriate targets for empirical research, this strategy fails to satisfy more subjectivist critiques.

Much environmental discourse, reflecting on itself, has adopted some of these 'middle' positions. Perceptions are seen to count. A person has interior space. The self as well as the outside world is viewed. The conception of individuals as meaning-creators, as Gordon has argued in his commentary on Du Bois's writings, is at odds with the view that they can best be viewed as playing roles in social groups or occuping positions in classes. In such perspectives, 'to know that role is to know all there is to know about the individual. In effect, there is no distinction between him or her and the social role ... The group, then, becomes pure exterior being. Its members are without "insides" or hidden spaces for interrogation. One thus counts for all' (2000: 275). The way people think about global environmental problems, their own actions, and about the reactions of governments and other actors to environmental issues, is central to much NGO activity. Environmental education – changing attitudes – is frequently ranked higher by their supporters than the strengthening of legal regimes. It is seen as a more effective, long-term strategy for combating ecologically destructive behaviour. There are echoes of the view in historical studies of changing cultural outlooks on wildlife and landscape in Europe and North America, and in searches for the connections between such views and the larger sets of metaphysical representations found in theological discourses.

Others are doubtful about the likely return on investment in the study of subjective worlds. Ecological and social systems, that is, are seen to be constituted of facts beyond the confines of mental products. They are complex and difficult to gauge in scientific research, perhaps, but offprints of objectivity nonetheless. From some sociological perspectives, the beliefs that individuals hold to may be little more than personal articulations of viewpoints embedded in social groups or classes. A related supposition is that behind such expressions can often be found the nuggets of interest. Southern critiques of environmentalism have represented it as hiding an anti-development agenda on the part of powerful countries and IGOs. Corporate affirmations in OECD countries of a conversion to green values have often been greeted at best with grudging approval by environmentalist critics. Apparent changes of environmental attitudes can also be interpreted as merely shifts of consumer taste. Greenery in this sense is a commodity like any other. Bacon suspected that habit rather than talk is the key to understanding such problems: people 'profess, protest, engage, give great words, and then do just as they have done before: as if they were dead images, and engines moved only by the wheels of custom'.

Discourses are important because they are central to the processes by which the world of social, and by extension environmental, things is constituted (Dant, 1991: 207). Two possible interpretations of these constituting processes are significant and also, in some forms at least, mutually contradictory. In the first, while it is reasonable to suppose that there exists a complex world of both social and ecological realities, and also that we are a part of it, the key point remains that the categories through which we comprehend such matters belong not to it but to ourselves. We have no way of standing outside the multiple worlds of discourses, that is, although we might be tempted to try to do this by elevating the stature of one, for example by deploying a particular criterion such as scientific method. And according to some critics, we cannot even rely on writers to provide more-or-less definitive views of the meanings of what they write: '– in the end how we understand what we read depends not on the private intentions of the writer but on the potentialities inherent in the public language in which he has chosen to write' (Harrison, 1999: 508).

Secondly, discourses can be viewed simultaneously as producers and the products of social forces. Gramscian enquiry begins with an acknowledgement of the way powerful groups in capitalist society produce and circulate convenient representations of social and economic realities, and goes on to emphasize the role of discourses generally in the structuring and transformation of social relations. This approach then takes us into a post-Marxist world. Discourses are important, not merely derivative; and they are seen to grow in specific spatial, cultural and temporal contexts (Torfing, 1998: 79–80, 93–4). Lacan, for example, classifies discourses according to instrumentalities, to the aims and actions of the agents with which they are associated: those of the university (for purposes of educating), the master (governing), the hysteric (protesting), and the analyst (revolutionizing)(quoted in Bracher, 1994: 107). These paths thus bring us back, in effect, to regions close to the first view. Both doubt the possibility of general or grand theory; and both, in a sense, resort to totalizing language to express their distaste for the totalizing tendencies of conventional social science.

Such processes are particularly evident at the junctures where discourses become transmuted into, or spring free from being, ideologies. Environmentalists' arguments have long been susceptible to codification in ideological form. This results partly from individuals' habits of thought and their cumulative responses to recurring events, but also, and more importantly, from practical considerations by environmental actors of the mechanics of political influence. An environmental group does not normally have to reflect from scratch about the significance of an oil spill or potentially hazardous food additives. The details of particular instances differ, but the general principles are taken as givens.

Such processes incorporate the reflexive gaze of discourses on questions of their own significance. The idea that ideas are important, perhaps crucial, formative influences on the way individuals and societies treat their natural environments is deeply entrenched in accounts of the long Western history of changing and multiple attitudes to nature. It is echoed in constructivist studies of the discourses of contemporary environmentalism (Macnaghten and Urry, 1998; Harré et al., 1999), and of a wide range of specific areas such as the social or cultural construction of acid precipitation, wetlands, moors and wildlife. The contrasting, though not necessarily contradictory, idea that ideas are embedded in social contexts is also part of the substantive content of discourses. Heller has suggested that this is a distinguishing feature of modern from pre-modern discourses, in that 'even if their ideas were embedded in their particular present to the same extent, they were unaware of this circumstance, whereas we are aware of it from the start' (1990: 37).

Praxiological discourses such as those on environmental problems, however, have to avoid being trapped into inaction by this kind of reasoning. Thus whereas an environmentalist account might attribute the stated views of an anti-environmentalist opponent to interest or political tactics, we can expect that it would want to draw, by contrast, a clear line connecting its own constituent propositions with the truths of an external reality. However, the idea that ideas cannot, even in principle, be evaluated according to their truth-values, raises so many difficulties that it is arguably incompatible with environmentalist beliefs. The absorption of subjectivist elements into these discourses carries the risk of environmentalist argument having to represent itself as lost in a relativistic sea of multiple interpretations. If discourses were in fact accepted as equal in this sense, the environmentalist might be unable to distinguish between the observations of, say, Greenpeace on tropical forests and Joyce on raspberry ram. The embedding of ontological and epistemological propositions in discourses prevents this from happening. In practice much environmental discourse retains a Lyotardian stamp of incredulity not towards modern but towards postmodern metanarratives.

The term discourse has developed different meanings, depending in part on disciplinary contexts. Environmental discourses nonetheless share a number of features with those of other domains.

Firstly, these form organized collections of arguments. Indeed discourse may be indistinguishable from argument. Mapstone sees this identity as a 'pervasive fact of social life', and defines argument as a rational search for truth, characterized by adversarial confrontation and rhetoric (1996: 219). Discourses thus contain mixtures of narratives, or accounts of facts and events, and consideration of options for action. Like any kind of communication, perhaps, they are designed to bring about some change in the reader or listener (Sornig, 1989: 95). Interwoven in these are accounts and rules that guide interpretations of the results of actions, the interests and motives of friends and foes, the bases for assessing the validity of certain beliefs, and the social and political impacts of specific discourses. These are typically collections, rather than a coherently organized whole. Environmentalist subdiscourses take place on toxic chemicals, endangered orchids, emissions-trading schemes for GHGs and many other topics. They take place in different places and times. There are compartments too for radical and reformist approaches, and for natural-science, critical-theory and other contending perspectives.

Secondly, particular discourses respond to the arguments and styles of others. Interactions define boundaries, though insiders and outsiders may disagree about where these are located. Environmental discourses typically encompass a broad range of commentary on social and economic matters, for example on globalization and the effects of trade liberalization, topics that some critics maintain transgress the proper boundaries of environmental discussion. External discourses can have significant impacts on internal argument-making: a closing off of receptors, for example, a toughened intransigence in the defence of core beliefs, engagement with a view to colonizing other discourse spaces, or proactive adaptation to differing perspectives through the importing of terms and propositions. Discourses on feminism and on development and North–South relations have been critical external contexts of environmentalism.

Thirdly, a discourse implies a sociological context of insiders and outsiders. Mechanisms provide members of discourse communities with definitions of who can participate, and play maintenance roles in keeping groups viable. Participation in any of the varied and self-consciously 'environmental' discourses of Western societies does not mean that ideological blinkers have to be worn, but it does entail a measure of acceptance and rejection of certain propositions. It would be incompatible, for example, with support for mining and other resource corporations having unrestricted access to protected areas. Partly this is achieved through consensus on the use of what has been called a 'stable core' of 'hot' words (Wagner et al., 1996: 332). In practice members of discourse communities use a wide range of cues. Environmentalists like to pride themselves on being skilled in the art of detecting the false (or other-discourse) uses of green vocabularies by companies as a cover for old-fangled corporate goals. Writing or reading the texts of discourses in this sociological context, moreover, has a self-creating function important for the identities and relations of NGO members.

Ancient traces

Discourses on natural environments have historically taken diverse forms. Their aetiology reflects old and still influential notions of the relations between individuals, society and nature, as well as recent constructs more characteristic of modernity. A core term of the environmental domain, 'nature' is rich in ambiguities, variously creative and obfuscating. The impediment makes it redundant as a tool for understanding environments. But, like class, it has the staying power of a resonant concept (Gerring, 1999: 371), and is thus a prerequisite for understanding environmental discourses.

Representations of nature have often been conflated with aspects of human transformative activities and the results of these. A lake killed by acid rain might nonetheless be admired on aesthetic grounds. Wilderness values may have to be diluted before they can be enjoyed by consumers who do not want to put up with the flies. The 'natural' landscapes of western Europe are the product of centuries of human transformative activities. The word may denote the living world with the human bits taken out, or both of these together with the physical settings of landscapes. In some usages it refers mainly to the environments of rural communities as opposed to those of cities, and in others to the physical laws that govern the workings of everything from expanding universes to collapsing Yorkshire puddings. Emerson, though immersed in transcendentalist conceptions of nature, was also influenced by developments in the study of nature defined in terms of empirical science (Wilson, 2000). Nature is either 'pretty scenery', according to Frost, or 'the Whole

Goddam Machinery' (quoted in Keith, 1980: 3). Nature personified appears variously as raw force, aesthetically pleasing prospect or, by way of overlapping concepts such as landscape, countryside or frontier, as a way of comprehending the human experience in a non-human world.

Nature language and conceptions of landscape, as in Austen's novels (Smith, 1994: 203), have also historically been a means of making moral and political judgements about societies and individuals. Typically, too, nature either contrasts with, or provides a set of clues for understanding, divine worlds. A sense of loss of these elements as a result of the rise of modern science has become a part of environmentalism (Haila, 1999: 48–9). On the one hand, nature's resources provide food, housing and clothing, and hence furnish Plato with the beginnings of the city state he describes in Book II of *The Republic* (though as an inveterate city-dweller he is as intent on rooting out the evils associated with urban expansion as he is to combat the appeal of the myth of rural bliss). On the other hand, aspects of contemporary environmentalism such as neo-Stoicism echo a later rediscovery of nature not only as a resource to meet human needs, but also as a receptacle of laws and principles that can be used as guides to help individuals, societies and governments flourish. Wisdom in Europe's Middle Ages meant living in accordance with such rules of nature, 'defined as an all-pervasive divine and fiery spirit that penetrated, animated, and unified the universe, and periodically consumed it' (Lapidge, 1988: 81).

Varying conceptions of nature were generated by later forces of economic and social change, and also played a part in shaping these. Somewhat like the notion of future panels of omniscient historians – or 'History' – looking back on and judging the present, recourse to views of nature as an arbiter of conflicting human claims has had enduring appeal. In the seventeenth and eighteenth centuries, nature, and the methods by which it was increasingly studied, cast light on the kinds of deals people make with each other to live in families and communities, the sources of war and civil strife, and the production of goods; in the nineteenth century, it became a silent ally of critics of the injustices of capitalist society. In Engels's account, societies move dialectically from a phase of harmonious living with nature, to a period of exploitation of its resources, culminating in an eventual return to a redefined cooperative relationship.

Outlooks on nature became formative influences in the development of nineteenth-century European and New World nationalisms. Early twentieth-century Australian identities were in part a reflection of images of the rural, constructed in a 'nostalgia for a lost, more isolated and primitive frontier, inhabited by rough and ready bushmen' (Waterhouse,

1999: 24). The sciences of nature often worked hand in hand with colonialism. Research in exotic locales, the setting up of museum collections, and the creation of protected landscapes, benefited from the power of colonial authorities while at the same time lending legitimacy to the pursuit of overseas territories (Drayton, 2000: 180–92). Approved concepts of nature were also built into the ideology of the Nazi regime. This produced in 1935 a law Lekan describes as the 'world's most comprehensive and stringent piece of nature protection legislation for its time'. Its significance, however, was more symbolic and legitimizing than real; failures to follow through with the kinds of conservation policy changes demanded by critics led to conflicts between important regions and the centre (1999: 384–6).

Diverse traces of nature thinking have left their marks on contemporary environmentalism in three ways. Firstly, nature-as-resource beliefs support much of the materialist side of environmentalist advocacy. Just as fears of the transience of the flow of good things from the land, rivers and oceans led to the invention of fertility rites and practical rules such as the rotation of crops, so they now foster sustainable-development arguments and policies. That these extend into debates about the structures of economies and the workings of governments also echoes traditional perceptions. If these institutions are faulty, so is the production and distribution of goods from resources. In *The State of War* Rousseau argues that the human need is 'not so much for men's care as for the earth's produce; and the earth produces more than enough to feed its inhabitants'.

Secondly, post-materialist arguments about the good life continue to draw on alternate conceptions of nature. Some are reflected in the strand of environmentalism that focuses not so much on economic sustainability, or which does so only tangentially, but rather on the ethical problems posed by nature, for example the existence and vulnerability of non-human species, particularly those outside cultural definitions of usefulness or risk to humans. Part of this perspective is grounded in sustainability criteria. Other arguments anticipate a transition to less dysfunctional economies as one of the consequences of ecospiritual change. Again, the idea has old roots. When a tenth-century Irish hermit-poet praised the 'delicious abundance' of eggs, honey, beer and other good things from nature, the point was, unlike Rousseau's, more theological than economic. When the resources of the Pacific north-west of the USA were discovered, exploitation was accompanied by the belief that Americans had a moral obligation to transform this gift of God (Bunting, 1997: 43). These kinds of views have led environmentalists to neglect what Easterbrook calls the 'case against nature' – that it is a major source of pollution and disease, and that through species extinction and other processes it is inherently self-destructive (1995: 140–5).

Thirdly, both of these views contain supporting metaphors of (im)balance or (dis)harmony to describe the problems societies have with their environments and the goals towards which they should be moving. Echoing Engels's dialectic of the alienation of humans from nature, Touraine writes of the 'history of modernity [as] the history of the slow but inevitable divorce between individual, society and nature' (1995: 153). Ecologically grounded writings argue that human–nature relations should nonetheless be seen more as a duality than as a growing estrangement of separate entities, since the structures and operations of economies remain subject to macroecological processes.

'Embiggening' environments

Many of the core terms and themes of environmental discourses can be identified with little difficulty. As with other conversations, however, words are used loosely. 'Ecology', for example, has increasingly referred not to an academic discipline but to the problems environmentalists like to talk about. Myerson and Rydin suggest that much of contemporary environmental discourse can be accounted for by the use of some thirty or so critical terms (1996: 230). Vocabulary politics is central to environmental politics. This responds to the creation and spread of terms like 'global warming' or the (enhanced) 'greenhouse effect'. The spread of biodiversity language in the 1990s reinvigorated a traditional area of environmentalism focusing on vegetation, landscapes and wildlife species. Lovelock's adoption of the literary term 'Gaia' to describe his views of the earth's complex dynamic systems has affected the differential reception of these among environmentalist and scientific audiences. We can look at these processes by taking account of three sets of forces. These are the mixtures of statics, expansive and other dynamics, and interactions with other discourses across boundaries, that structure and bolster the contours of environmental discourse space.

Statics

The topics that make up environmental discourses vary over time and across cultures. Environmental NGOs characteristically differ over priority issues. Some topics central to contemporary Western environmentalist accounts – air and river pollution, the consequences of human population growth, the provision of clean water for communities, the transformative power of technologies, the destruction of habitats – were often discussed in the past with the aid of different vocabularies. Labelling and regrouping these as 'environmental' highlights continuities from the past, but blurs understanding of the specific contexts in which each arose.

Internal fragmentation in discourses makes it uncertain whether there is indeed a coherent whole here. An environmentalist response would be that there is, but the simple answer is undermined by the range of subjects and conflicts finding their way into discourses. Princen argues that three different views of the environment shape social and political responses to issues. He describes these as the frontier view, where the task is simply to exploit resources with minimal or no interference from government or legal regimes; environmental protection, where the environment is viewed as a source of amenities or raw materials, and primarily economic calculations have to be made about their use; and sustainability, which deals with the interconnections among economic and biological systems and compels attention to a wide range of social, economic and political decisions (1998: 396–400). A few of the more familiar and ruggedly persistent themes around which argument flourishes can be identified briefly.

The most elementary, and historically the most enduring, of these has been the focus on the consequences of human population growth. 'The most significant event in the organic history of the Earth in the last 100,000 years has been the rise in the world population of human beings in the past two centuries' (Darling, 1970: 7). The observation allows environmentalists to draw on a range of related propositions from ecology. Though explicit attention to population growth and distribution questions varies over time and in different discourses, it nonetheless provides a foundation for environmentalist commentary on many social and economic problems.

Traditionally in Western states from the 1960s these have often been identified as stemming from pollution. The term covers questions as varied as local issues of effluents from factories and agrochemical run-off from farms, to large-scale questions of the planetary climate-change consequences of carbon-based energy economies. The focus captures important and urgent problems, but overlooks others. Cities, chemical plants or highways can be environmentally clean in the sense of producing minimal pollution, but they still leave ecological footprints. These include the effects of the energy processes that maintain them and the opportunity cost of lost habitat. A second focal point of environmentalism, especially from the early 1980s, centres on problems of sustainability. The term, which can mean many things, forges an alliance with natural-resource issues of forests, fisheries and soils, and also with larger issues of the management of economies. It thus connects environmental discourse with a broad spectrum of economic questions on the agendas of Western governments and IGOs. Resource-based questions of environmentalism have often turned out to be less tractable than many of those associated with pollution. The knowledge bases for understanding critical processes, such as those underlying the collapse of the cod fishery off Canada's east coast in the late 1980s, have been weaker. A third large subdomain of mainstream environmentalism concerns the values, non-consumptive as well as economic, of biological diversity. Part of this set of issues derives from the continuing influence of the focus on the protection of wildlife species and natural habitats that marked conservation politics in the late nineteenth century. Another more utilitarian strand accommodates consideration of the pharmacological, agricultural, ecotourism and other benefits of maintaining biodiversity for sustainable economies.

While such discourses ostensibly focus on particular sets of questions – climate change, biodiversity, noise pollution, and so on – they do so characteristically by appealing to general propositions about human beings and their societies, ecological processes, and the nature of knowledge. Even science-based environmental discourses are thus implicitly also exercises in the social sciences. They accordingly make use of general-theory templates that serve to interpret complex social and ecological situations.

Central among these are interpretations and applications of ecological principles and laws. The anchor-issue of population growth, just noted, is crucial because of its connections with both resource questions, and hence sustainability discourses, and with larger normative critiques of Western society, and hence with radical environmentalism. As Schlesinger has written, 'There are few basic axioms of ecology, but one of the most fundamental is that which predicts the ultimate collapse of a population showing exponential growth in a closed environment' (1991: 349). The proposition opens up discussion of complex relations among demographic, social and economic variables. The wide circulation of UN demographic forecasts has helped to keep population arguments at the core of environmental discourses.

A closely related mix of arguments surrounds normative consideration of the relations of humanity and nature. Differing strands of environmentalist reasoning rest comfortably on the apparently contradictory propositions that these two big entities are either separate or intertwined. McLaughlin argues that 'the dualistic sense of nature' is ever present. 'Just as we are surely within nature, we are also, just as surely, a distinct sort of creature, not to be simply identified with "all that is"' (1993: 5). The distinction leads to different forms of environmentalism. Soule contrasts 'those who want to save the world to save the people' and 'those who want to save the world to save the world' (quoted in Beazley, 2001). The 'within' argument, associated for example with deep ecology, takes to task many other areas of environmental discourse for looking at the world through species-centric lenses. The 'outside' argument maintains the priority of human interests. It defends sustainable-development strategies and precautionary thinking on the grounds that these serve human needs better than do older practices of resource exploitation. The latter extends too to consideration of the needs of imagined future generations, which I will come back to in the next chapter.

Both viewpoints also contain propositions about human nature, or more specifically views of either cooperation or of conflict as the principal characteristic of social relations. The former belief tends to find more receptive environmentalist audiences. Representations of pasts characterized by nature–human harmony, non-Western contemporary cultures held to embody this virtue, and alternative futures engineered to recapture it, have at times been influential in environmental debates. Related images of 'politics' shape views of environmental problems. Environmentalist notions of the political have shaped concepts such as that of the commons at the global level, or support for practices such as round-table dispute-resolution procedures at local levels. Environmental, or ecologically enhanced, armed conflicts thus tend to be explained not so much by human conflict-proneness as by the dysfunctional nature of relations between populations, economies and resources.

The arguments of Western environmental NGOs have also put emphasis on individual choice and power. The argument contains a mixture of empirical elements (individual acts aggregate to societal-level change), ethical considerations (individuals have a moral duty to act in certain ways), and pragmatic and mobilizing concerns (the more embedded the belief in individual efficacy, the greater the political and cultural influence of environmentalism). The proposition draws on traditional civic values of participation, particularly in the context of US environmentalist discourses, though it tends to leave unresolved many questions of the optimal forms of polities. Many approaches to these issues coexist in environmental discourses. Individualist environmentalism tends to have a special affinity, however, with civil-society models. This is a result of traditional squeamishness about embracing the state for fear of co-optation, and more specific concerns about the social-policy gaps left in the wake of the retreat of the neoliberal state.

A further divide relates to the purposes of the individual's engagement with environmental questions. While this can be seen as reflecting an Aristotelian desire to improve the polity and the lives of its citizens, it can also be interpreted in more Epicurean terms as part of a hankering for self-development. The two overlap in arguments about the values of identity-through-community produced by environmentalist engagement. However, the latter emphasizes the ecospiritual and other non-consumptive rewards of environmental activity, is disinclined to judge such activities by the criteria of social change or even material benefit, and is more prone to see politics, whether conventional or radical, as inherently corrupting. From liberal-democratic perspectives, such beliefs seem to be associated more with narcissism than environmental problem-solving (Walker, 1998: 848). Hence, perhaps, the source of the puzzle that, while environmentalist beliefs may be widely held in a culture, or widely transmitted through communications media, they are not necessarily translatable into an equivalent force of citizen participation in environmental groups.

Finally, epistemological debates of various kinds lie behind environmental discourses. There is some tension between the view of environmentalism as either a body of empirical, normative and widely applicable generalizations, or as little more than a signpost directing enquiry towards the particular. The latter view reflects a classical notion of ethics as aimed at 'local rather than universal knowledge' and as speaking to 'issues of the moment, problems that bear upon a group of people in a specific space and time' (Miller, 1996: 233). Arguments about the place of science are also central. Good science, variously defined, is taken by some as the appropriate basis for environmental knowledge and, consequently, for sound public policy. Statements about global warming have been criticized on such criteria as lacking adequate data, and as prone to measurement errors (for example, through use of terrestrial and urban-area monitoring stations), weak inference and lack of enquiry into alternate explanations of data, and misinterpretation of causal relationships. Other environmental critics question reliance on conventional scientific methodologies. These are variously argued to produce environmentalisms that are too cautious when it comes to action, dismissive of traditional knowledges and biologies (Medin and Atran, 1999), prone to embrace technological solutions, neglectful of contributions from the social sciences and humanities, or insufficiently self-reflective as observers of their own cultural relativism.

Environmental discourse is also characterized by metaphors: of the earth as a spaceship, in Boulding's arguments from economics in the 1960s, or, in Lovelock's Gaia, as a kind of self-regulating physiological organism. The choices of these are significant for the structuration of environmentalist argument. As Myerson and Rydin put it, 'metaphors are argumentative: if the Earth is an organism, it is not a spaceship' (1996: 149). Many such images have shaped environmental accounts: the greenhouse in climate-change discourse, for example; balance and harmony metaphors in depictions of the relations between cultures, populations and environments; and the language of disease, parasitism, or symbiosis to describe the human presence on the planet. Three metaphors in particular are worth noting.

The first is the idea of an interconnected world, and of holistic thinking as the way to divine its workings. This is a productive device for uncovering hidden links, for example between aerosol-spray use and ozone-layer loss in the 1970s and 1980s, or for advising prudence and recall of the precautionary principle before embarking on economic and technological steps that might have unforeseen consequences. It is also used to challenge vertical (agency- or department-based) as opposed to horizontal (cross-sector) divisions of environmental tasks inside governments. By emphasizing the intricacies of connectedness, however, it tends to discourage ranking exercises aimed at setting priorities among competing environmental policy goals (or even attempts to distinguish these from 'non-environmental' questions). Such exercises are an integral part of government. As noted in the previous chapter, the metaphor meets occasional criticism from professional ecologists, though it fits comfortably with scientific images of the biosphere as 'in fact one single inter-connected unit' (Patten, 1991: 291). The recasting of environmental argument as a quest for the wholly frail is also objectionable to critics who would prefer to rejig it on a more postmodern, and less totalizing, basis.

A second structural metaphor is that of problem-solving. This is so fundamental to environmental discourse that it is almost subversive to isolate and comment on it. Related forms of problem-solution thinking have been identified as a common feature of discourses in housing, health and other public-policy areas. As in environmental-policy debates, approaches to the definitions of problems and generative metaphors of the nature of these set up discourses oriented around a search for solutions (Rochefort and Cobb, 1994). The motivating assumption is not that this is a simple task – that 'solution' can be easily 'clapped on problem' like 'a snuffer on a candle', as Beckett puts it (quoted in Gibson, 1996: 23) – but rather that keeping an eye on bottom-line action requirements is a trusty guide through the thickets of complexities we should expect to encounter.

The sense of urgency typically linked with searches for solutions reveals a third metaphor, that of depth. While the imagery is widespread, as in the radical dismissal of proposed environmental-policy reforms for only 'scratching the surface' of problems, it is a more structural feature of deep-ecology and related ecocentric critiques and some forms of critical-theory discourses. In the first of these, much environmental commentary is criticized as anthropocentric or culture-dependent. The word 'ecological' in these contexts tends to have richer connotations of depth and structurality than does the term 'environmental'. In the second, the structural properties of economies, or aspects of discourse processes, are put forward as the key to understanding ecological deterioration. The metaphor often finds its way into depictions of environmental problems in the language of crisis. The field of political ecology has been defined as beginning at the point where 'our ecological crisis is taken as given' (Atkinson, 1991: 4). Partly because of its resonance in critiques of the contradictions of late capitalism, interpretations of Marxist ecology are also attracted to this terminology (O'Connor, 1998: 135–43).

Dynamics

Changes of environmental themes and vocabularies have been integral to environmental discourses. Some accounts are too exuberant to be contained within conventional public-policy frameworks. Concepts, myths and styles of argument disperse to colonize other habitats such as international trade, and meet strong resistance when they land there. In the word coined by Jebediah Springfield, clay-footed town-founder from *The Simpsons*, the environmental imagination is thus continually being 'embiggened'. As one environmentalist has described his own redefining efforts, 'It ceased being just trees and clean air and water. Now, environment includes economic equity and economic justice and human-rights issues' (*The Economist*, 27 September, 1997: 30). Redefinitions of the environmental extend to issues of war and peace, gender, theology, economic development and, through metaconcepts such as risk, to the multitude of stresses confronting people in advanced industrialized societies.

Mission creep at some point prompts questions about the boundaries of environmental discourse. Unchecked, environmentalism can transform itself into a simulation of the physicist's elusive theory of everything. A comprehensive ecologically grounded account of the nature and malfunctioning of economies and societies ceases to identify distinctively environmentalist problems, since nothing is left that is non-environmental. The approach accordingly loses much explanatory power. Pragmatic considerations also periodically enforce retrenchment and contraction in discourses. Inside governments, the environment is defined in discrete and circumscribed ways. There is no neat match between the specific responsibilities of environment departments or agencies on the one hand, and the definitions of the scope of the environmental by external critics on the other. Groups attempting to influence governments accordingly have to focus lobbying efforts on specific issues and conventionally accepted notions of environmental questions. Similar constraints limit larger exercises by NGOs in environmental education. These also compete with other groups in contested areas, for example in the area of trade and the environment. The resulting interactions often focus attention more on the limited range of applicability, rather than the universality, of environmental critiques. Expansion processes have none the less taken environmental discourses in several innovative directions, including earth perspectives, social contexts, globality and time.

Firstly, some environmental accounts have set issues in the larger earth-systems perspectives discussed in the previous chapter. These extend the range of inventories of the environmental to phenomena such as earthquakes, vulcanism, continental drift, glaciation and other aspects of the earth's physical environment. This is partly an orienting or heuristic process in environmental discourse. It is also partly a consequence of the rise of the anthropogenic puzzle in environmental-policy debates. A feature of climate-change debates in the 1980s and early 1990s was the problem, often highly politicized, of disentangling evidence of the 'natural' forces of global change, such as the historical patterns of rising temperatures in interglacial periods or naturally occurring methane production, from the influences of variables such as increased GHG emissions from cars, factories and farms. While from one perspective it is the anthropogenic hypothesis that is typically 'environmentalist', other viewpoints incorporate a much wider range of earth-systems phenomena as integral to the wider social-ecological project. The approach can be extended almost indefinitely to incorporate into discourses very large-scale atmosphere–ocean cycles or solar-system processes. Environmentalism contains, however, its own limiting conditions. Discourses, after all, are oriented to action. The possibility of catastrophic hazards from cometary impacts has been discussed by writers since Laplace in the 1790s (D'Hondt, 1998: 158). The topic squeezes in on action criteria in some environmental conversations, but discussion of the implications of, say, the predictable stages of the life cycle of the sun and the eventual death of the earth does not (except, perhaps, as a metaphysical parenthesis or warning against hubris). This brings us back to the predicament of the observer, who faces the choice between characterizing as 'environmental' only those questions defined as such by social and political actors, or reaching for a more autonomous stance that describes their environments in terms that might not be of their choosing.

A second process of expansion of environmental discourse takes questions deeper into social and economic contexts. Issues such as the international transport of hazardous materials, or the creation of national parks in areas where people live, require political decisions inseparable from issues of justice. Any decision with environmental consequences, whether made by individuals and households or states and multinational companies (MNCs), entails costs and benefits the distribution of which is unlikely to be equitable. Much of the agenda of environmental discourse can be redefined in these terms.

Thirdly, the global dimensions of problems have become increasingly integral to environmental accounts. The familiar rationale is that appreciation of this bigger picture effectively grounds and motivates actions at local levels, the two levels being treated as aspects of the 'glocal'. A focus on globality blended with study of economic forces highlights a crucial region of environmental argument. It draws attention to what Cox describes as the macro-cleavage between 'those peoples and social groups that have become structured hierarchically into the global economy from those that have become marginalized and excluded' (1997: 248). Environmental issues often take substantially different forms in the economic circumstances of societies of the North and South. Sometimes, though, NGO emphasis on global environmental problems produces an unanticipated consequence in Western countries. This is that problems elsewhere – tigers in India, Japanese 'scientific' whaling, deforestation in Brazil, dam-construction in China, forest fires in Indonesia – can become in a sense more urgent and tangible for environmentalist sympathizers than more chronic, and typically less flamboyant, problems of environment-related poverty in Northern city neighbourhoods or declining rural communities.

A focus in discourse on the global also opens up dynamics that problematize the state. The issue links efficacy and community concerns: that parochially nationalist solutions to environmental problems are likely to be ineffective, and also that the problems confronting individuals and civil-society groups in one country are widely shared elsewhere. Both the global and local levels, then, are seen as challenges to the normative claims, if not in the short term the power, of states. Yet the need for and the efficacy of states in solving environmental problems remains a central issue in debates. To the degree that economic globalization is represented as threatening the capacities of polities, cultures and civil societies, more robust environmental nationalisms that can institute sanctions against international norm violators, and use greater regulatory authority against domestic backsliders, attract more interest and political support. The identification of environmentalism with globality creates other difficulties. It may imply a universalizing claim that the definitions of environmental problems commonly found in contemporary Western cultures have wider applicability. In practice, this proposition is not so much a static feature of discourse as a basis for the development of contending interpretations of the global. Environmental discourses often contain counter-points that reflexively label their own arguments as cultural products, and which warn against their casual transport to different economic and social contexts.

Reflections on changes over time constitute a fourth mechanism of growth in environmental discourses. Temporal factors make an appearance in these in various ways. Many questions have a prehistory with origins predating the formation of environmental movements, for example the concerns voiced in the 1840s about the polluting effect of French railways (Lyotard, 1993 [1989]: 96). Environmentalism has itself gone through several discernible phases since the late nineteenth century. Discourses occasionally reflect, even if they do not explicitly reflect on, these continuities. Accounts drawn to earth-system perspectives likewise include some notion of geological history. Memories, and representations of these, as Said has put it, 'touch very significantly upon questions of identity, of nationalism, of power and authority' (2000: 176). However, consideration of time is also often displaced to the margins of environmental discourses, or into academic specialisms such as environmental history. Part of the vigour of environmental critiques of modernity and its costs rests on their capacity to communicate a presentist vision dependent on a forgetting or highly selective reinterpreting of past events: it is in the present that the urgent problems arise, that is, and with them the just-around-the-corner possibilities of effecting decisive change.

At stake too is the issue of environmental progress. Neither pollution nor actions to contain it are by-products exclusively of advanced industrialized societies. In England, for example, the Fleet river in London became impassable as early as the 1650s because of dumping. A history of environmentally related stench, poverty, violence and disease persisted well into the nineteenth century, as towns failed to provide such goods as adequate burial places or alternatives to the supply of drinking water from communal butts (Chinn, 1995: 90–1). The health consequences of lack of basic facilities for clean water and for waste disposal were observed by Engels in the 1840s. The smogs of the early 1950s caused thousands of deaths. There is thus ample evidence of change in policies and attitudes, and of the capacities of societies to learn from crises and to alter energy-use, waste-disposal and other practices. This leads some critics prematurely to the conclusion that most environmental problems either are being or soon will be solved as the 'age of pollution' ends (Easterbrook, 1995).

Transboundary processes

Discourses occupy social niches. They are also characteristically associated with particular places. Cresswell writes that both aspects are connected with the creation of difference: 'An outsider is not just someone literally from another location but someone who is existentially removed from the milieu of "our" place – someone who doesn't know the rules' (1996: 154). Environmental representations are shaped in part in processes of interactions, tacit or otherwise, with those of other discourses. These include those of various business, libertarian, consumerist, traditional Marxist and other observers of environmentalism. Nature can be oversold, some critics of the Canadian literature of the 1980s found: 'We were desperate for stories about hamburgers, subways, television, anything but bloody loons' (Smith, 1998: D10). Some of these exchanges harden attitudes on both sides. Discourses become more monologic in form, in that the aim of listening is to 'develop better arguments *against* an interlocutor's point of view' rather than to accommodate, integrate or build bridges (Tuler, 1996: 61).

Part of the problem is that environmentalist notions, as we saw earlier, can happily lodge in a variety of political contexts. Neoliberal, socialist and other broad frameworks typically incorporate views of environmental topics. To the extent that the cultures of US politics can be divided into orthodox and progressive strains of ethical beliefs and policy perspectives (Jensen, 1997), environmental arguments of various kinds straddle both camps. This proneness to manipulation suggests some limitations to environmentalism as either a reformist or radical project. Saul, for example, has argued that western political elites have traditionally organized societies around sets of answers requiring their expertise and legitimizing their roles and power. Even if environmentalism were to be entrenched in these societies, he suggests, it would become just another set of answers and would not itself therefore transform political and economic structures (1992: 3–4, 315).

Interactions with animal-rights, feminist and development discourses raise other problems. Animal-rights discourses overlap with those of environmentalism's endangered-species concerns, but define different priorities, for example the treatment of animals in agriculture or critiques of trapping. During the 1990s these questions were in some ways more successful in mobilizing agents than environmentalism. There were interstate consequences too, for example in the EU's relations with Russia and Canada.

Ecofeminism, criticized in some of its earlier forms for essentializing conceptions of women and nature, and hence for failing to promote emancipatory goals, has changed and expanded. It has been described as no longer oriented around earth ethics, but concerned with 'domination itself, in all its forms – whites over people of colour, men over women, adults over children, rich nations over the Third World, humans over animals and nature' (Van Gelder, quoted in Vautier, 1998: 159–60). This expansive dynamic, however, sometimes blurs the distinctiveness of ecofeminist critiques. It also risks universalizing ecofeminist concerns and in so doing appropriating the voices of environmental activism in the South, in part, according to Sturgeon, by treating indigenous women in such movements as idealized figures (1999: 257–61).

Finally, the propositions and styles of argument of Western environmentalism have interacted in various ways with arguments about nature associated with Southern places. In her study of political conflicts surrounding a biosphere reserve in a region of forest in the southern Yucatan peninsula of Mexico, Haenn (1999: 477–80, 488–9) emphasizes the importance of differences in the ways ecological systems are given meaning. The creation of the reserve in 1989 led to strong opposition from local farmers. They redefined the environmental-conservation objectives lying behind the reserve in terms of class conflict, issues of the ownership of natural resources, and critiques of the role of the state. The politics of reserves of this kind, she argues, stem from proactive strategies by governments, international agencies and transnational NGOs to change the prevailing ideas of local ethnoecologies and attitudes and tilt these towards those of the largely urban world of environmental conservation planning.

In many economic-development contexts, environmentalist arguments thus interact with others through a variety of social, cultural and political mechanisms in which self-consciously 'environmentalist' representations may not circulate widely. As in the case of the Brazilian Amazon, social constructions of environmental problems in Southern societies may thus be responses to processes whereby external actors impose views on members of local communities (Ciccantell, 1999: 293–4).

Guinness describes community responses to the construction during the mid-1990s of a flood-wall in Yogyakarta as a response to a severe flood of 1984 (1999: 101–6). While local residents initially approved the project on flood-control grounds, the project soon became redefined in a variety of other ways. In particular, it created a mixture of fears and hopes with regard to land occupancy and rights of ownership, for example on the part of those residents who had traditionally built houses on sand-flats by means of the successive creation and filling in of fishponds. Incentives to change ownership patterns and land use also followed from anticipation of the prospects of significant increases in commercial land values if the project succeeded. Attitudes were also shaped during the process of construction, for example in light of poor work by some local building contractors, growing resentment of the presence and demands for meals by armed forces personnel engaged in the work, and a lack of opportunities to express complaints about the effects of the project on wells. Guinness concludes that the project shows 'the contradictory impact of development projects. Constructed by the state with local residents' welfare in mind, and largely welcomed by them, the wall has raised community tensions and undermined the strength of local community' (1999: 106).

Constructions and global ecology

The discourses of environmentalism reveal both diversity and also a core of environmental orthodoxy in prevailing Northern accounts. Stories of these topics have historically been embedded in Western cultures. Academic accounts, for example in social anthropology, take up related themes of the complex links between societies and economies and their natural environments. Study of the ways individuals construct social and environmental worlds, and of the metaphors and other mechanisms that structure such processes, is also a central feature of attempts to conceptualize agency. A notion of agency is inapplicable if an individual or social actor is considered merely as acting in response to environmental perturbations, without having some picture of these and expectations about the results of possible actions. Representations of the environment have inherently expansive tendencies, especially where these interact with themes from related discourses on gender, human rights or development. The process carries the risk of environmental accounts explaining too little because they try to do too much. A related risk is of enquiry being stunted by the packaging or commodification of environmental ideas into communicable sets of answers.

Divergent methodological tendencies characterize approaches to such questions. One strategy is to treat representations as objective facts, and to study their formation in socialization and political processes and their effects in terms of influences on behaviour and interactions. This approach, while useful in terms of understanding voting behaviour or the interactions of multiple players in relation to specific environmental issues, leaves open some important questions. It can assume too easily, for example, that the constructs through which people latch on to aspects of their natural environments correspond with 'objective' features of these. The observation of discourses, whether by academic observers or by participants in them, also raises questions about the observer's role and effects: for example, whether what is 'environmental' or 'ecological' is what members of discourse communities say they are, or whether these and other more specific definitions can be superimposed on accounts from the outside. To the extent that actions are reflexively interpreted through discourses, moreover, environmentalists could also be caught up in a paradox of inaction resulting from the absorption of cultural criticisms of notions of out-there realities. Much environmental discourse tends for such reasons to be more modernist than postmodernist in character, and reluctant to eject the ontologies inseparable from accounts of the nature of global environmental problems. Approaches by way of critical subjectivity nonetheless make creative contacts with environmentalist themes through a shared emphasis on the local and the particular.

5
Duty Rosters

A feature of much discussion of environmental matters is that questions are often framed in ethical terms. This is taken as natural, as an obviously appropriate way to proceed, rather than as the result of a careful choice of intellectual strategy. Instead of moral reasoning being thought a part of a larger problem – as we might say there are ethical dimensions to issues of health care, gambling laws or unemployment – this aspect has often become virtually synonymous with, or at least a big part of the foundational structure of, environmental enquiry. Discourses on environmental topics in IR, though, tend to be an exception. An ethically grounded view of environments is consistent with either a traditional utilitarian definition of ethics as pursuit of the optimal spread of happiness among members of a society, or a simple ethical injunction to act prudently. Yet environmentalist discourse often seems to go further. The normative impulse insinuates notions of right and wrong, duties and rights, and care for others into regions of discourse and social action that can just as plausibly be viewed as practical matters of material self-interest. And the 'others' towards which moral obligations are defined are varied. They include non-human, non-mammalian and non-living entities, as well as abstractions such as ecosystems and, in the case of future generations, people who do not exist.

How useful – or how obligatory – is it to say that such things as polluting factories or the genetic manipulation of agricultural crops are ethical questions? Does the global context of enquiry alter approaches to this problem? The first part of the chapter reviews some of the problems of the approach. Following this, the respective criteria sets of the functioning of ecological systems and of human needs are considered, in part through reference to issues of biodiversity and endangered species. Central to both is the issue of whether environmentalism requires some kind of cosmopolitan social organization that transcends traditional state-based forms.

Applied ethics

We could speculate on the reasons why environmental discourse has taken on this ethical coloration. Since *ought*-statements cannot be derived from those of the *is*-variety, moral obligations towards the environment cannot be drawn from empirical observations assessing ecological deterioration. The phenomenon stems perhaps from the moralizing temptations facing countermovements such as environmentalism. Resort to moral-high-ground arguments can be a useful political tactic, though it can backfire. As discussed in the previous chapter, there is also a history of environmental questions being defined in terms of big, if not very useful, concepts such as nature; and some of these traditionally bring in theological issues that connect with arguments about the supposed sources of moral principles. Further, academic philosophers have played important contributing roles in the development of environmental discourses, and brought with them perspectives on the moral bases of human actions common in their discipline. And environmental issues do make contact at many points with traditional ethical issues, such as consideration of the consequences of one's actions for the well-being of others, or of the appropriate treatment of other species.

Biological, emotivist and interest-based views of ethics are among those that suggest that such options are probably misguided. The first finds explanations of moral beliefs in biological evolution. As Campbell points out, though, questions of the truth-value of such beliefs are logically different from questions about their origin (1996: 30). In the second we find 'the version of relativism that says that ethical beliefs are not the sorts of claims that are capable of being true or false, but are merely expressions of emotion or "pro-feeling", and that these emotions and pro-feelings may be either individual preferences or culturally transmitted dogmas' (Ellis, 1998: 29). Thirdly, statements of value preferences are seen to reveal only the interests of actors, and the nature of the power relations among them. As Thrasymachus told Socrates in *The Republic*, these include the interests of powerful groups in society. Privileged classes, as Niebuhr observed, typically resort to moralizing to protect their interests. The argument logically extends to the strategies of the weak too, though defences of their interests have often been more normatively constructed by sympathizers.

These questions have a long heritage in commentaries on international relations. States are conventionally seen in IR to deploy moral arguments when this suits their interests. Failure to recognize this truth, it is argued, leads states, and non-state actors, into trouble. The use of moral principles thus becomes subordinate to factors 'as pedestrian and lacklustre as self-interest or patriotism' (Gross, 1993: 50). In classical, or at least realist, IR, different sets of values characterize the relations of individuals on the one hand, and those of states on the other. The two do not mix. We cannot satisfactorily transfer ethical rules from the former (the injunction against killing people, for example) to the latter.

Further, integral to the Westphalian division of the world into sovereign states is the principle of non-intervention. This traditionally required the simultaneous acknowledgement of differences of domestic moral habits among states and their governments, and rejection of these as for the most part not pertinent to diplomacy, or at least to the determination of foreign-policy objectives. Altruism in this context is compatible with, and may require, hostility towards persons beyond state boundaries or minorities within. Ethical rules are thus primarily designed by insiders with a specific political community, such as the nation, in mind. Even ostensibly universalist rules include definitions of the circumstances in which they can be broken, bent or ignored in dealings with outsiders. This stance is defensible in some accounts because of the absence or weakness of normative foundations of world order. Realism can thus be seen as a form of consequentialist ethics in conditions of anarchy, a way of resolving the inevitable problems that deontological ethics confront when we move outside the group into the uncertainties and hazards of international relations (Brilmayer, 1999: 192–3).

The influence of this tradition of thought in IR has carried over, to some extent, into debates on the environment. Wapner points out that IR writers have produced a large and growing literature on international environmental law and politics, but adds that 'it is curious that they have yet to concentrate on the ethical dimension.' Scholars 'have generally failed to raise, in a sustained manner, issues of moral duty, principled understandings of appropriate conduct, or simply the challenge of moral deliberation as it relates to the causes of and responses to transnational environmental issues' (1997: 213).

Three sets of factors are reversing this relative neglect. Firstly, many, though not all, environmental arguments, like those in domains such as human rights or economic development, challenge normative as well as empirical planks in the realist-IR platform. Analyses of international environmental regimes are weakened to the degree that they share a traditional IR disposition for marginalizing domestic politics. Secondly, and more generally, the dual circumstances of post-cold-war (or perhaps post-Westphalian) politics and economic globalization direct attention towards transnational interactions among social and ethnic groups. Through such instruments as immigration policies, citizenship rules and taxation powers, states retain much of the capacity to shape definitions of political community along conventional lines, but this capacity is increasingly challenged by transnational developments. These have also attracted growing interest on the part of political theorists, whose concerns have traditionally been centred around problems within states. Thirdly, and especially in this more complex world, an understanding of society and politics requires a more hermeneutic stance in which actors' own understandings of the world are analytically central. These understandings are in part normatively oriented. Through what Bilmes calls normative talk, 'we praise, condemn, evaluate, and do various of the other acts that constitute the social process' (1986: 161).

But the question remains whether, or in what sense, specifically environmental talk does deal with ethical issues. Ethical issues are certainly much in evidence in discourses. But are these located in a normative reality acknowledged by (non-environmentalist) others, or is the ethical framework better understood in a functional sense as something riveted on to a motor driven by different power-sources? Before engaging with issues in the arguments themselves, we should note the relative ease with which environmental arguments can be transformed into exercises in morality.

Sustainability, frequently talked about in ethical terms, can in many ways be regarded more as a 'better mouse-trap' code than a solution to a moral predicament. It is prudent to manage a fishery so there are some fish left for future years, or to control car or factory emissions before these cause an increased incidence of upper-respiratory disease. There are, however, many alternative views of sustainable development. Some versions are denounced on ethical grounds simply because they are designed to promote a more effectively operating capitalism. Radical accounts, Davidson (2000) argues, contain significant components since they address critical questions of social justice and thus reach far beyond conventional ideas about the running of capitalist economies. The sustainability issue takes on more of a recognizably ethical slant, perhaps, (1) if the foresight and long-term thinking required to arrive at these conclusions is assumed to constitute a higher form of reasoning that can usefully be labelled ethical (for example, because it considers the needs of future generations, or because the reverse posture, for example neglect of the consequences for human health of economic activity, is taken to be unethical); (2) if it can be seen to reflect a utilitarian calculation of the preferences and concerns of all relevant actors, and an extended assessment of the likely consequences of all or most possible actions; or (3) if conflicting interests and values are seen to call for a theoretical resolution in the form of the articulation of general principles of right behaviour. Some arguments for sustainable economies, while valid on other criteria as alternative means towards generally accepted ends, tend to weaken if the ground shifts too far towards ethics. But if they take an instrumentalist route, they become vulnerable to normative complaint on other grounds. The Brundtland Commission's concept of sustainable development has been criticized as excessively oriented towards traditional goals of economic growth, despite its protestations of concern for issues of social justice and democratic participation (Langhelle, 1999: 138–40).

Further, conventional definitions of utilitarianism suggest a posture of neutrality as regards the content of people's preferences. If the principle is interpreted to mean that one person's preference for enjoying wilderness, non-GM foods or clean air is aggregated equally, somehow, along with another's for, say, watching soap operas or soccer, environmentalism may come out the loser. So this minimalist approach will not work for the environmentalist. Further, societies have other goals than environmentalist ones. Reducing pollution may result in unemployment or lost market opportunities. While environmentalists have their own ways of balancing the costs and benefits, others may disagree. The federal tax authority in Canada defined the issue in these terms in 1999 when, in rejecting the application of Greenpeace to have an associated foundation granted charitable status, it argued that it could not be assumed 'that remedying any and all forms of pollution always conveys a public benefit' (*National Post*, 5 June 1999: 1).

Forms of practical means–end thinking run through other ethical-environmental arguments. Arguments about duties towards wildlife species are made partly on grounds of their usefulness in forestry, agriculture and pharmacological research, and partly on criteria that define protection as morally obligatory regardless of the value (or threat) of these species to humans. Yet criteria of use and interest are difficult to avoid even in the latter argument. Practical benefits may take the form of scientific knowledge, spiritual growth and self-development, or ecotourism company profits. Duties to animals are thus subsumed under notions of inter-personal ethics. Arbitrary distinctions between wild and domesticated species of animals traditionally shape the culturally based ethical rules governing eating practices (Eder, 1996: 85). Many nineteenth-century middle-class girls in the USA were taught vegetarianism to prepare them for the moral role in adulthood of preventing alcohol and sexual abuse by males (Nelson, 1994: 13).

We can look further into these questions by distinguishing the moral obligations that are held to arise from the natural world, and those that are defined explicitly in relation to human needs.

Nature's claims

In his story of a dog, Kafka writes that 'all that I cared for was the race of dogs, that and nothing else. For what is there actually except our own species? . . . All knowledge, the totality of all questions and all answers, is contained in the dog.'

The first of these perspectives criticizes as irredeemably biased those environmental positions that take their criteria from definitions of human values. Midgley, for example, is critical of conventional sustainability arguments. Her reservation, however, is not that these are too oriented towards the manipulation of economies in the direction of traditional economic-growth goals, but rather that such arguments (and presumably by extension the views of social-justice critics of sustainable development too) are based on considerations of human needs. They thus reflect a 'species-egoist approach' to ethical problems of the environment (1997: 92–3). They are arguments from Kafka's dog, before it began to observe its own reasoning. However, while critiques of anthropocentric bias in moral beliefs about the environment aim to make these vulnerable to the charge that they are (merely) expressions of self-interest, they tend to leave unfilled the freshly opened-up ground that requires criteria other than human needs in order to lay the foundations of an alternative ethical argument about nature.

One traditional way around this kind of problem in environmental discourses has been to bring on board criteria from other philosophical or theological constructs. The argument is still that there are moral obligations to nature, for example to non-human species, but that these exist separately from any notion of their capacity to contribute to the satisfying of human needs (or, perhaps, that the satisfying of human needs is a consequence, whether anticipated or not, of fulfilling these other first-order duties). These strands are often inseparable. Appeals to the needs of nature are at some level statements about human interests. This blending recurs in discussions of the principle of reverence for nature, for example in the accounts of the risks to the earth posed by technology developed in Martinson's novels in Sweden in the 1930s and 1940s. There are similar tensions in the feminist notion of care, defined as 'a species activity that includes everything that we do to maintain, continue, and repair our "world" so that we can live in it as well as possible' (Tronto, 1996: 142). The definition of moral obligation, that is, is inextricably tied up with calculations of the material and non-consumptive benefits we can reasonably expect to obtain by acting altruistically towards nature. Isolating these – carrying out the duty without expecting the reward – is possible in some interpretations of deep ecology, but the advice does not travel well.

Expectation of benefit is also present in redefined forms in more explicitly theological arguments, for example in discussions of the application to global environmental management of the Hindu concept of *satyagraha* (Dwivedi, 1994: 98–101; Egri, 1999: 63–6). The 1990s saw significant expansions of the links among the world's religions on environmental-conservation questions. Some of these traditions, however, seem in practice to be tolerant of bad models of environmental practice. Buddhist and Taoist thought, for example, have been influential in shaping Western environmental discourses. However, interpretations of their applicability have 'often been suffused in the West with an idealized glow' that overlooks environmentally harmful practices in Asian societies (Clarke, 1997: 178). In Japan, Shinto and Buddhist beliefs underpin philosophies of respect for nature, but there has been widespread cultural acceptance of such problems as pollution, excessive packaging, unsustainable logging practices, harmful activities by companies in Southeast Asia, and extensive habitat destruction, as well as complacency about the lack of public and scientific knowledge required to tackle these issues (Callicott, 1994: 102). Religious frameworks also bring extra-environmental considerations into the evaluations of behaviours and options. Foltz cites the example of an old river festival in Bangladesh that celebrates a connectedness with nature's forces, but which is open to criticism from Islam on the grounds that it contains polytheistic elements (2000: 64). In environmental discourse in the Christian tradition, advocacy of stewardship of nature, as opposed to the notion of dominion over its resources more familiar to its critics, aims to avoid these costs, but is still arguably anchored in a notion of ultimate human benefit.

Debates on duties towards other species are central to many of these regions of environmental discourse. These overlap with animal-rights debates, for example through extensions of concepts of rights-possession to members of other species, but are distinguishable from them. The moral obligation of protection or stewardship can be defended on human-needs as well as ecocentric grounds, for example, in the aretaic argument (Bond, 1996: 136–7), by defining such behaviour as part of good human character. But ecocentric lines of argument hit difficult terrain when critics point to an obligation to privilege human needs. They tend to deflect this attack by attributing it to a false nature–human dualism. Naess (1999), for example, has argued strongly against the proposition that an ethical stance derived from deep ecology is indifferent to human suffering.

Some notions of duties towards other species are weakened, however, to the extent that species can be recognized as the artefacts of biological taxonomists. 'In the Darwinian flow of life, the sharp lines between species are our own creations. At the margins, species can overlap either in the historical development of one out of another or in the extreme divergence of individuals from the norm' (Lachs, 1997: 299). In wider public debates in Northern states, moreover, species attributes such as size, intelligence and complexity of social organization have traditionally pressed the claims of whales, orang-utans, elephants and other spectacular megafauna (regardless of the country they happen to be located in). Non-mammalian or non-vertebrate species are less favoured by a sense of the importance of evolutionary proximity or by anthropomorphizing imagery. This has been a traditional constraint on the activities of environmental NGOs promoting conservation of spiders, snakes and amphibians in Northern states. Dolphins caught in tuna-fishing nets become an 'environmentalist' issue, but the tuna do not (except perhaps in the resource-management sense of environmentalism). Traditional uses, myths and perceptions of species as threats to personal security or economic livelihoods, also shape ethical calculations. As Cooper writes, 'few people could claim, hand on heart, to view the prospect of a cockroach-less world with the same gloom as they do that of a world without, say, elephants and dolphins' (1999: 232). Few people, similarly, would respond to save-the-world questions with calculations about the impacts of humans on the rest of the biosphere that might lead to advocacy of a people-less, or even a significantly less peopled, world. The integration of ethical, religious and use perspectives on fauna and flora tends to be more characteristic of non-Western or pre-modern cultures.

Some traditional notions of ethics make these approaches problematic. Personhood, of both the recipient and the producer of good deeds, is conventionally a crucial, though not the only, criterion for judging ethical actions (Rovane, 1998: 233–4). The attribution of personhood to a member of another species, common though the anthropomorphic tendency might be, is difficult to defend. A condition of ethical reciprocity is thus absent. Further, most ethical criteria sets would establish a stratified system of rules that would put obligations towards humans universally ahead of those towards others. Even so, it can still be argued that members of other species have some claim to be considered as having moral standing. The qualified claim, that is, does not imply that they have a claim to moral equality with humans.

Further sets of ethical choices are made in practice among species. The 'God committee' listing US endangered species makes such choices routinely (Rolston, 1993). In endangered-species calculations, value goes up according to rarity or probability of extinction. Or, more specifically, moral obligation is seen to intensify to the degree that human activities threaten a species. However, the magnitude of global habitat transformation produced by human activities is now such that the distinction between anthropogenically induced change on the one hand, and 'natural' extinction processes on the other, is increasingly difficult to sustain. The obligation to conserve the habitats on which the survival of wildlife species depends thus becomes greater as the forces of globalization grow. Endangered species could be, and are, protected in zoos or gene banks, but this minimal strategy is inadequate; even national parks and protected areas have been equated with prisons by their more severe critics.

This definition of the ethical problem, however, leads logically to the requirement of defining moral obligations towards a large group of abstract things such as types of ecological zones, or towards physical things such as the rocks and chemicals that constitute the bases of these (Rolston, 1988: chs 4, 5). This raises even more acutely the issues of personhood and reciprocity. For some critics, the holistic reasoning such arguments are based on could even reflect a stance of arrogance. Seeing nature as an interconnected whole has been criticized by Scriven for coming 'dangerously close to a general picture of nature that makes it out to be a human artifact, namely, a machine' (1997: 180).

The more bio- or ecocentric the ethical posture, then, the more questions of human social and political organization become second-order requirements. The complexities of these can also be ignored. Identifying this kind of duty towards nature implies after all a global viewpoint. It does not matter, or affect the interpretation of the obligation, where a species resides or a habitat is located. The awkward facts of political boundaries – those of states, for example, or ethnic and religious identities – become morally irrelevant or at most secondary concerns. If the government of a country objects to external environmental pressure to preserve on its territory a particular species or habitat, outsiders may thus be tempted to acknowledge this pragmatically only as a temporarily annoying obstruction. Recognizing these practical (and normative) realities of a territorially organized world of polities, however, blunts the force of the original ethical argument. This brings us to the second category of problems.

Redefining human needs

One theme of the discussion so far has been that assessments of ethically appropriate behaviour towards nature easily slip over into rationales for the pursuit of self-interest, or a wider species interest. We can now turn to ethical frameworks in which this is largely a non-issue. The varied dimensions of environmentalism – conservation of natural resources for sustainable use, the protection of individuals and communities against the hazards of pollution, and so on – are in this second image defined in terms of human needs. Environmental questions are here normative because they are directly related to questions of health, living the good life in a thriving economy, removing injustice, and post-materialist self-development. They are associated with deaths from pesticide use in the South, illnesses from the avoidable contamination of water supplies, and the ethical dilemmas resulting from biotechnological advances in the North. Environmental ethics attempts to explain individual and community impoverishment in late modernity, and to propose remedies. The question of duty towards nature is thus altered. Instead we have to ask: in what ways does the goal of cultivating ethical relationships *among humans* require obligations towards nature? The latter, that is, now recede into a second-order requirement or a means towards an end. Tied to these questions, moreover, are issues of good environmental governance.

Assessments of environmental problems touch at many points on traditional ethical issues. They include questions of individual autonomy and community. Davradou and Wood argue that the former value is degraded, and with it the quality of life in communities, if ecological conditions deteriorate significantly. If humans are 'hounded by the continuous need to satisfy the basic requisites of life' they cannot plan for the future. In such scenarios, 'social instability, impending famine, drought, and war force them to make choices guided by fear' (2000: 74). But if the content of ethical rules is left open, or if they are left to be determined in processes of discourse among humans, they might even cease to demand our attention at all. The way communities are defined is a central feature of this normative problematic. The idea of a person as a traditional rights-possessor often remains, for example in assertions that individuals have a right to clean air, but the bleaker notion of an isolated self disappears. The environmentalist conception is probably closer to the Japanese way of seeing individuals in the context of their mutual dependence in communities.

Nature is thus normatively significant in terms of the resources it provides for the benefits – non-consumptive as well as material – of human societies. The ethical issue relates in part to the manner in which we make use of these goods: through market principles, for example, government coercion, or systems of small-scale communal practices, or by way of rules of thumb such as the precautionary principle or the argument that common property should be used in such a way that enough is left for others. Natural habitats can thus be changed, transformed or even obliterated. The enhancement of river habitats, for example by the construction of weirs, is acceptable if such practices can be shown to increase fish populations, while avoiding the costs of fish-farming such as the risks of introducing disease or genetic change among wild populations. Economic practices are open to ethical challenges if the costs and benefits arising from nature's transformation by human ingenuity are characterized by significant inequalities – if it can be shown, for example, that disease rates resulting from air or water pollution are spread unevenly among income groups, or that the siting of hazardous-waste sites discriminates against racial minorities, or that the social cost of dams, as in India, is disproportionately borne by those compelled to move. Practices are also exposed to normative complaints on different ideological grounds, for example if environmental regulations are seen to be harmful to property principles or to threaten the social well-being that is seen to result from economic growth.

Such normative critiques of cultural or economic practices are thus different from those of the nature-based image. Firstly, the origin of moral obligations towards nature does not lie in assumptions of the rights that a member of a species might have, but rather in the normative discourses on rights and duties among humans in different cultures that define obligations towards these and other features of the natural environment (Frost, 1999: 602).

Secondly, the fact that selfishness can be observed in behaviour that actors claim to be ethical is not necessarily relevant. In classical economics terms, the individual pursuit of self-interest serves to create, rather than undermine, ethical communities. Thiele has argued that selfishness is integral to ethical postures in a variety of ways. Ethical discourses can be viewed from an evolutionary perspective, for example, as products of natural selection (although this does not, as Campbell reminded us earlier, negate the truth-value of ethical statements). Both social-justice and economic-development sustainability arguments are thus products of selfish human drives. However, they nonetheless constitute ethical choices, in Thiele's view, since they are designed to ensure the long-term viability of human economies and cultures (1999: 11, 27).

Thirdly, the argument that there are in nature entities, such as wildlife species, that have rights or other claims to our moral attention – separately from any features we autonomously impose on them – is historically associated in Western traditions with a contemplative perspective. In Thoreau, nature contributes too to material well-being, but ideally does so only at the level of isolated individuals or very small communities. The urgent problems of large modern economies are from a human-needs perspective largely ignored in ecocentric complaints. Such societies may be dismissed outright as founded on bad ecological principles. Further, nature tends to be seen from the Thoreauvian position primarily as an aid to self-development. From a liberal-democratic perspective this looks suspiciously like a concern for self-cultivation rather than for the building of civil society (Walker, 1998: 848).

Interest has thus underscored a number of supposedly normative viewpoints on environmental questions. Criticisms like this are part of the games of environmental politics. Conservative or neoliberal critics routinely insist they can detect self-interest – or pressure from 'special interests' – coursing through the activities of environmental NGOs or government environmental agencies. Environmental groups respond with similar scepticism to the pro-environment arguments of timber or agrochemical companies. And some defences of nature conservation have traditionally been associated with stakes in other institutions. These include conservative values of the countryside in England, and traditional hunting rights in the case of migratory birds in parts of France and Italy. Concepts of property were transferred to colonies in the nineteenth century, and took the form of defences of wildlife, the creation of national parks, and, as with earlier land enclosures in Europe, sanctions against the new offence of poaching. Yet arguments about conservation are not all reducible to considerations of interest. Various definitions of temporal, territorial and political communities are interwoven through these debates.

Problems of temporal communities

Central to the claim that environmentalist argument has a significant ethical component are expressions of concern for the welfare of future generations. The imagined environmental community typically spreads spatially to become global; here it extends over time. The magnitude of the costs future generations will incur as a result of present-generation failings accentuates the problem of intergenerational justice. McMichael argues that this is 'the first time, at a global level, that one generation has conferred a *negative* legacy upon future generations', and that this 'poses an unprecedented moral problem' (1993: xv). Global climate change in particular focuses attention on the ethical problem of the distribution of costs and benefits of economic activities across generations (Page, 1999: 53–4). Azar and Holmberg (1995) have argued that this climate-change debt in terms of GHG emissions from contemporary unsustainable economies can be quantified. Such analyses are not conventionally part of economics. Or, as Vercelli puts it, in orthodox economic theory 'time is fully reversible, exactly as in classical mechanics from which it drew inspiration' (1999: 79).

The view that this is an ethical matter is based on the premise that members of future generations can be counted as others to whom ethical obligations are due. Goodin discusses the analogy of saving. An individual saving for the future, he argues, is likewise acting altruistically, since the person of the future is a different person from the one doing the saving (1992: 76). More generally, the notion of a duty to members of future generations resembles not so much saving-for-self, as leaving behind inheritable property for family members.

However, several problems arise. Presentist concerns inevitably underlie the argument that as-yet non-existent persons are entitled to a say in contemporary decision-making. There are, or will be, many such generations. A call for full intergenerational equity, achieved for example by way of a temporal extension of Rawls's original-position argument, would thus be impossible to effect (unless, perhaps, a short-term time-limit is built into the ethical calculation). In such arguments, moreover, these multiple generations tend to blur into a single cross-temporal cohort. This is often constructed as a like-minded constituency, one more uniformly environmentally aware, and less pluralistically conflicted, than our own. Its members seem to have clearer-sighted views of the rightness of certain kinds of environmental actions and the wrongness of much of the way in which contemporary economies are structured. Definition of the obligation is complicated by the possibility that they may be the lucky survivors of past-generation environmentalist advocacy of the need to limit their numbers by population control, or that they are the results of past-generation decisions without which they would not exist (Page, 1999: 54). The approach also suggests, perhaps, a heavy-handed social-engineering project designed to manipulate temporally distant cultures.

A redefined proposition might be that, since the value orientations of any specific future generation cannot be known in advance, its members should be provided with a range of choices predicated on the implementation of sustainability in the present. Vercelli prefers a criterion by which development 'may be considered sustainable only when future generations are guaranteed a set of options at least as wide as that possessed by the current generation' (1999: 83). Like Parliament in the Westminster model of government, that is, no generation could then make substantive decisions binding on its successors (though each would presumably act within the constraints of a constitution decreeing sustainability practices). But we might even so become complicit in enforcing early-2000s mind-sets on future others who have different values, or even, by the simple fact of ensuring they have adequate resources, in facilitating choices by future-generation communities of options such as fascism or plunder. Sustainability arguments attempt indirectly to deal with such hypothetical problems in two ways: by an Enlightenment posture that assumes the ethical and material progress of societies, or by defining sustainability as necessarily intertwined with democratic participation and social justice. Built into the sustainable-development project, then, are implicit assumptions about the causal relationships among environmental protection, democracy and economic development (Langhelle, 1999).

Problems of spatial communities

This brings us back to the territorial as opposed to the temporal extension of notions of community. Environmentalism has at times been divided on the respective merits of global, national or other conceptions of community.

The values of nationalism and ethnicity continue to influence environmental discourses. The definition of the nation or the ethnic group is in a sense an extension of the kind of moral deals made by members of any group. Contracts establish obligations towards insiders, and rules about dealings with outsiders that may exclude these from the ethical considerations applying inside the group (Bond, 1996: 210–11). National values have also influenced and reflected attitudes towards natural resources and biodiversity. In the IR-realist argument the former constitute a significant element in national power; their conservation and sound use is thus crucial to the exercise of power politics and effective statecraft. Biodiversity arguments can be extended to other aspects of national identity. Parts of France's endangered-species legislation, for example, traditionally highlighted the importance of national as opposed to other European species of wildlife; and foreign species introduced into Britain in past centuries, such as pheasants, rabbits and fallow deer, have gradually seeped into definitions of national heritage (Boardman, 2001). Assessments in Northern cultures of the value of nature in the South reflect other ethnocentric values. Further, the state apparatus attached to nations becomes for NGOs a useful tool for imposing sanctions on outsiders who violate insider codes, as Japanese whaling ships do those of US environmentalists. The associated critique of WTO-supervised globalization is thus not that it thwarts progress towards global civil society, though that secondary objection may be raised, but rather that it restricts national capacities to design national environmental codes.

However, ethical reasoning on environmental matters has traditionally been more expansive than this. It has tried to integrate the concerns of the local and the global. A neo-Stoic strand has pressed forward the claims of an imagined global community as the relevant ethical context for framing the environmental actions of communities. This approach does not sit comfortably with traditional IR notions. According to Bull in the 1970s, if doctrines of the protection of human rights in international law become entrenched, '[t]he way is left open for the subversion of the society of sovereign states on behalf of the alternative organizing principle of a cosmopolitan community'. Similarly on environmental matters, the state system should not in his view be dismissed as dysfunctional. States have both the resources for effective action, and also stronger normative claims to be representative actors than NGOs (1977: 152, 294–5). They can also if they wish, or if their citizens insist, opt to participate in collaborative or global schemes. Rawls likewise is reluctant to take the step towards the cosmopolitanism that some see as the natural organizational form to protect human rights and other values. He recognizes, though, a duty of 'well-ordered peoples' to assist 'burdened societies', that is, those that 'lack the political and cultural traditions, the human capital and know-how, and, often the material and technological resources needed to be well-ordered' (1999: 82–3, 106).

Various forms of cosmopolitanism, however, tend to spring up if we begin by focusing on the well-being of individuals rather than on structures such as states, and proceed by rejecting the automatic coupling of these in normative views of the state (Barry, 1999). Only in these kinds of ways can many of the issues of global and transborder ethics – duties towards non-patrials, for example, or the reconciling of deontological ethical rules with the parochial codes of nations and ethnic groups – be satisfactorily handled. Environmental discourses have thus explored alternative forms of governance. In practice some arrangements grow naturally out of immediate cross-border environmental problems, for example those among states within the USA or the local transborder administrative arrangements of German state governments. The many possible routes range from reform proposals for tinkering with international institutions, through Kantian prospects of growing international government, to advocacy of global community-building through ethical principles.

Another option, though, is simply to disconnect ethical issues from consideration of alternative models of governance, and to deal with environmental policy problems through existing institutions and processes. This strategy reflects environmentalism's pragmatic biases. Indeed, the fact that environmentalism often seems more compatible with cosmopolitan alternatives is in part a consequence, paradoxically, of the way its ethical debates often de-ethicize awkward problems. The ethical stance, that is, dissolves environmental sub-questions into a variety of problem-solution tasks. Technical expertise may be required to deal with these, but, having determined metaproblems and the rationales for action, ethical discourse can be pushed aside. Pragmatism, whether liberal or radical, thus becomes the appropriate ethical response. The fruits of environmental ethics 'must be directed towards the practical resolution of environmental problems – environmental ethics cannot remain mired in long-running theoretic debates in an attempt to achieve philosophical certainty' (Light and Katz, 1996: 1–2).

This strategy is transnational or global in nature, then, because of the spatial expansiveness of environmental problem-solution logics. Canada or Sweden cannot deal with problems of acid rain without bringing, respectively, the USA or Britain into transborder regimes. Managing domestic environmental problems thus becomes a way to establish the international credibility that helps secure the transborder cooperation that reinforces the efficacy of domestic measures. No country acting alone can mitigate problems of anthropogenic climate change. The globalist argument thus becomes less about the pros and cons of constituting ethical communities beyond the borders of states, and more about the practical requirements of environmental policy. These requirements may include transnational forms of governance based on redefined notions of political community, but if they do it is only because these can be shown to be pragmatic means towards achieving environmental goals. Environmentalist discourses implicitly make contact in these ways with functionalist approaches to governance. These will be taken up in chapter 7.

Problems of political communities

Forms of social and political organization at national, global and other levels vary considerably according to democratic criteria. Accounts of the requirements of sustainable economies conventionally incorporate provisions for broad participation in governance activities by civil-society groups. However, the history of environmental thinking is less consistent. When the urgency of the environmental 'crisis' has seemed too overwhelming, more directive or coercive options for the exercise of state authority have been urged. Future crises could herald a return of such advocacy. Even appeals to democratic values have diverse implications. How should decisions on environmental matters be made within ethical communities, however these are defined? Is there an ideal ecological polity, or forms of these at global and national levels that are more ecologically benign than others?

An initial problem is that polities are not monochrome. They are not set up or maintained solely or even primarily for the purpose of solving environmental problems. And, even if, hypothetically, a polity were to pursue the aim of existing in ecological harmony with its surroundings, there would still remain both a multitude of interpretations of how this could best be achieved and a wide array of conflicts about the choice of specific environmental objectives.

Secondly, then, ecological ideas are associated with diverse traditions in the history of Western political thought. Environmentalism is not a stand-alone political creed. Like nationalism, it gains much political vigour from this claim. But, also like nationalism, in environmentalism a lot of the big picture of what societies, including global societies, should look like is either left blank or filled in with taken-for-granted assumptions (Freeden, 1998). Links have to be made with other sets of values – socialism, neoliberalism or feminism, for example – before these empty spaces of the environmental canvas can be sketched in. It is often these connecting tissues that enter political debates and that shape attitudes towards environmental issues. Thus an economic conservative may see in environmentalism an attack on property and market principles, and a socialist a failure to understand the character of class relations. Nationalists may suspect a cosmopolitan-elite plot. Conservative critics in the USA have portrayed UNESCO's World Heritage network, and national conservation initiatives such as the Heritage Rivers system, in this light. Interpretations of environmentalism also adapt to changing circumstances. As liberal environmentalism responded in the 1990s to the retreat of the state, it tended to be increasingly sympathetic to non-regulatory approaches to environmental policy, and to civil-society voluntarism as the engine of environmental change.

Thirdly, democracy is notoriously an inexact term. Thus an environmentalist who wants to be a democrat can do this and still justify use of the term in many ways: by being a liberal, for example, a radical activist, an eco-warrior, a one-worldist cosmopolitan, a federalist, or an advocate of decision-making by scientific experts. Imaginary polities organized on the basis of optimally 'democratic' environmental principles would vary correspondingly.

Environmentalism nonetheless retains an air of being contrary. It likes to see itself as a challenger of conventional economic wisdoms, a thorn in the sides of comfortable elites and their clients, a visionary wired into alternative futures. Its ideas are available for use in broad critiques of modernity, and in attacks on the inadequacies of societal structures to control the costs of unchecked economic-growth goals. Yet environmentalism has insinuated itself into the lives of modern polities in diverse ways. In France, ecological argument was notably absent from the civil protests resulting from the 1960s student movement. But these events precipitated the rise of French environmentalism, and gave it its antistatist posture. Yet the French state has 'transformed itself into a colossal green Leviathan, its politicians and bureaucrats propounding environmental laws, decrees, regulations, and guidelines faster than the body politic can absorb them' (Bess, 2000: 6–9). Environmentalism, however insipidly from the point of view of more radical critics, has become institutionalized in Western governments and international regimes. It has a long history of association with conflicting political beliefs, including those of conservative defenders of old ways. Advocates of change include both cautious incrementalists, inside and outside governments, and others arguing that environmental problems require, and are precipitating, social transformation at domestic and global levels.

Moreover, the character of the governance problem varies with different polities and cultures. Tuler has defined the liberal-democratic problem in this way: 'A fundamental challenge of a democratic society is to manage environmental policy disputes that include significant value conflict in a manner that protects the rights of individuals, but simultaneously provides "voice" to diverse values and interests and is responsive to collective needs and values' (1996: 1–2). Guides to appropriate processes leave open the critical issues of who is to participate in the making of environmental decisions, or the extent to which coercion of dissenting individuals (or of offending states beyond a political community's borders) is justified. The ethical problems change, moreover, to the extent that states under conditions of globalization are being displaced by more powerful economic actors. Paehlke has argued that a commitment to democratic values has characterized contemporary environmentalism, and has distinguished it from older approaches that were content to see the right decisions made by trained officials (1996: 18–19). Jurisdictions vary, however, in the scope they allow for participation, in their responses to the environmental-governance claims of aboriginal organizations, or in their provision of tools such as courts. The latter were commonly used in the USA in the 1970s in citizen suits against corporate polluters, but the avenue became more restricted in the 1990s. Round-table and other exercises in reconciling stakeholder views have been a growing practice in Western societies. Politicized or ideologically based views of the criteria for participation weaken these. In British Columbia's forest politics, issues of relations with hunters' organizations led to conflicts in the 1990s among nature-conservation and other environmental groups (Wilson, 1998: 62).

A possible long-term ethical strategy in such situations focuses on the instrumentalities of discourses. Among other things, Baumslag writes, these generate the knowledge of the experiences of others that is central to the art of treating them ethically. This can be done, he argues, through use of novels and various rhetorical methods (2000: 132–3). Normative approaches drawn from Habermasian and other practices of discourse ethics have a particular resonance with environmental problems. Avio (1999: 526–7) argues that there are similarities, or complementarities, between these approaches and those derived from the study of institutional law and economics. In both, that is, persons are encouraged through deeper understandings of their own situations and those of others to reach accommodation with opponents. Apart from instances such as Burton's work on conflict communication, however, notions of interest at the core of traditional IR have made comparable approaches in that field rarer. Governance debates there point rather towards interest-based or plurality models of cooperation. Risse has argued nonetheless that there is scope for application in IR of the concept of communicative action, that is, according to Habermas, 'when the action orientations of the participating actors are not coordinated via egocentric calculations of success, but through acts of understanding'. Such action points to 'an argumentation in which participants justify their validity claims in front of an ideally extended audience [and] presuppose the possibility of an ideal community "within" their real social situation' (quoted in Risse, 2000: 9–10).

Such a notion of community need not lead to a cosmopolitan or globalist conception in which intergroup and territorial boundaries lose significance for actors. It would be compatible in principle, for example, with a model in which state actors through their representatives communicate in a fashion subsumed within a reasoned consensus about the interstate society of which they are members. For global environmental matters, though (and probably for many others), typically built as these are around the actions and preferences of multiple types of actors, particularly civil-society groups, this approach has to be refashioned to accommodate an extended social complexity embracing diverse players. Moreover, actors play multiple games. Interwoven through arguments about good citizenship are down-to-earth factors of interest, for example on the part of companies resistant to GHG-emissions regulations or householders who object to the extension of habitat or species-protection laws to their properties. Complex multiple-actor negotiations carried out on the basis of interest suggest different models of governance from those of communicative action. These approaches are nonetheless compatible with domestic round-table and related approaches to environmental dispute-resolution, on the one hand, and with functionalist and other conceptions of global governance on the other.

Two sets of problems – conceptions of the state, and the views individuals have of moral rules – persist in attempts to make governance arrangements more responsive to principles of environmental ethics. First, there have traditionally been conflicts among civil-society groups on core questions of their relations with governments and on the roles of these. While much environmentalism has tended to espouse a civil-society ethic, and while there has been increasing support for voluntarism and government-NGO partnership schemes on the part of OECD governments, there are risks in taking national governments too far out of the mix. Civil-society groups lack the resources to monitor and enforce good environmental standards, as often do municipal governments. The other side of this predicament is that, when embedded in government, environmentalism risks provoking political backlashes. There was mounting conservative criticism of environmental regulations in the USA in the early 1980s and again in the late 1990s.

Secondly, individuals' support for environmental rules may wane if these are viewed as restricting the choices of everyday life, as time-consuming, as possibly leading to increased taxes, or as forces that replace enjoyment with earnestness and chores. To take a different example, lowering speed limits reduces highway fatalities and pollution, but people find the rules inconvenient, and so arguably are prepared to trade lives and poor air-quality for headache-avoidance (Norcross, 1998). The environmental rules in question may be societal or take the form of government regulations. Whether or not these are perceived as ethically based, they are vulnerable to variations in the metarules individuals have about the nature of their obligation towards rules. Simpson distinguishes between internalists and externalists on this issue. For the former, moral rules are a motivating force, readily translatable into action. For the latter, however,

> believing one is under a moral obligation is connected with desiring to fulfil it only if one also happens to feel sympathy towards those to whom one has the obligation, or is concerned about the impact of law or custom upon persons who neglect their obligations, or has a general desire to fulfil whatever one is obliged to do, or has some other motive for compliance extrinsic to the belief in question.
>
> (Simpson, 1999: 201–2)

Like a latter-day temperance movement, environmentalism has sometimes appeared to conceal if not zealotry, then at least an excessive taste for shaping people's daily lives and thoughts in morally uplifting ways. Ethical metarules, then, suggest circumstances in which persons can allow themselves to ignore moral obligations such as those towards the environment, for example by enjoying things like music instead (Wilson, 1993: 283). These classes of rules also include devices for judging the social or geographical distance between individuals, and the consequences of this for interpreting rules. Normally we might expect the sense of obligation to decline with distance. On some environmental questions, though, distance may intensify it, as with some Western environmental NGO debates on endangered fauna in Southern countries.

Ethics and global ecology

Though less centrally so than some of its claims would have us believe, ethical discourse takes us into many critical regions of global ecological enquiry. This is despite its vulnerability to the criticism that the problems it addresses – for example, responses to other species or extrapolations of sustainable-development notions to specific cases – reflect more a basis in interest, biology, prudence or plumbing. Discourses organized around ethical reasoning also open up a wide range of questions of governance. Normative argument is fundamental to arguments about the organization of polities, the relations between rich and poor societies, individual agency and community, and other issues directly relevant to environmental enquiry. Empirically too, the ethical beliefs of actors have consequences that can be studied. As ethical issues generally have become more central in recent years in IR, they form a fertile ground for exchanges among environmentalist and other observers of trends in global society.

An implicit cosmopolitanism in some environmentalist accounts is nonetheless problematic. Conventionally in IR, states, and by extension other large global actors, do not have moral characters in the way individuals do. Cosmopolitan notions of transboundary political community can have paradoxical or self-defeating consequences. They may be associated with advocacy of ethically questionable forms of intervention in other societies on ecological, human-rights or other grounds. Yet closely related notions are often central to environmentalist arguments, for example in accounts of moral obligations to future generations. Such arguments do not normally restrict their scope to communities within a narrowly defined territorial area. Normative arguments may treat questions of sovereignty, nationalism and ethnic identity as second-order questions, subservient to the supposedly higher moral imperative of a protectionist stance towards nature that implies the ethical acceptability, or even the duty, of interventionist strategies. However, environmental discourses, especially those shaped in part in interactions with development discourses, usually also contain self-regulating normative restraints. They tend to be critical, for example, of environmentalisms neglectful of traditional knowledges or biologies as potentially complicit in forms of cultural imperialism in the South.

6
Minute Circumstances

The concept of the individual, like the related one of the person, is notoriously diverse. Individuals are the entities which possess rights in liberal democratic theory, and are among those which, in radical conceptions, become energized as sites of resistance to oppression. They observe themselves and their experiences and actions. They are variously viewed as discrete beings, or as more blurry parts of wholes – good ones like communities or social movements, as well as bad ones like totalitarian states. They pursue their interests and make rational calculations if an economist is watching, and trim their sails according to role expectations if a sociologist, or at least a more traditional one, takes her place.

Of the multiple forms of methodological individualism, those centred respectively around rational-choice perspectives derived primarily from economics, and theoretical work in sociology on individual agency, form contrasting – though in some ways, I will argue, complementary – approaches relevant to problems in global environmental enquiry. While approaches to agency and rational expectations have been applied to multiple levels of social organization, this chapter focuses for the most part on individuals, and on these perspectives as ways of understanding the relations of individuals with local and global environmental problems.

Individuals, acts, outcomes

Seeing the world as stocked with individuals, each endowed with deep internal space, is a convenient device. It is a way of rooting complex social phenomena. 'Attributing mental states to a complex system (such as a human being) is by far the easiest way of understanding it' (Baron-Cohen, 1995: 21). Further, if we can interpret someone's action as being rational, it becomes 'understandable' (Goldthorpe, 1998: 185). We even see computers as quasi-persons, while keeping separate the knowledge that they are not; the disposition can be manipulated by computer designers (Fogg and Nass, 1997: 552). The representations constructed by individuals from groups or from states may have something of this personalizing quality. Plato developed the point formally in comparing his ideal polity with an individual (or, rather, in likening the state 'to an individual likened to a state' [Provencal, 1997: 76]). In analyses of international relations, however, describing 'nations as if they were human beings' is viewed as an oversimplification that robs frameworks of explanatory power (Schokkaert and Eyckmans, 1999: 204).

Putting individuals at the centre of social theory also gives analysis both a realistic touch, and a sense, of which sociologists particularly have been fond, of connecting with the lives of ordinary people. Thus while we still need larger social-theory concepts, Flew argues, we should 'recognize, and never forget, that all social activities, and the operations of all social institutions, are and cannot but be the actions of individual human beings' (1991: 65). Giddens is among the sociologists who have grounded their analyses and theory-building enterprises in concepts of agency, and who have insisted too that this notion can apply only at the level of the individual.

A focus on individuals has increasingly become an important vehicle for comprehending globality, just as globalization has been recognized as a critical dimension of the lives of individuals. The first of these points has some counter-intuitive elements because of the distance and disparities of scale and power between the individual on the one hand, and global economic, social and political structures on the other. Structurationist concepts of the ways in which individuals constitute the structures of their societies have nonetheless been adapted by sociologists to incorporate extensions to global processes.

The second proposition has been a growing influence on sociological thought. Historically, Albrow has argued, Western concepts of individualism were shaped in part by indigenous ideas and circumstances, but also, because of colonialism, 'as much by the need to comprehend otherness' (1997: 33). Postcolonial notions of selfhood have also been formed in complex transnational contexts. Through these, for example in the writings of Naipaul, we see that 'the creation of a self carries great cost. It involves a separation from others, from the world of home, that can never be entirely complete' (Gorra, 1997: 95). Globalization exerts many limiting – and, its supporters would emphasise, emancipatory – effects on individuals. These were evident to different nineteenth-century observers, from defenders of the peace-and-prosperity benefits of free trade to advocates of the socialist transformation, nationalist struggle or pursuit of aesthetics held to be required to combat the globalizing excesses of modernity. The points are echoed in early-2000s criticisms of the costs of globalization for communities, environments, government social programmes, local economies and the psychological well-being of individuals. The process also has its defenders, and not only on classical economic grounds. Beck's qualified optimism leads him to argue that in a globalizing modernity, older certainties are 'being replaced, if we are fortunate, by legally sanctioned individualism for everyone' (1998: 28).

However, modernity also prompts questions about the usefulness of seeing the world as most comprehensible through theories that nudge us towards individuals. One type of objection is not to methodological individualism as such, but rather to the social-science frameworks used to capture it. The complaint is made variously by historians, political theorists and ethnomethodologists in sociology. An individual, that is, is complex and elusive. In using the concept in social or economic analysis, this complexity has to be disentangled. The various routes to simplification, for example the abstracted individual in economics who consumes and works or in psychoanalysis who projects and represses, make us drift too far in this view away from real individuals. An opposed criticism is that societies and economies are also complex, and that these have properties that cannot be fathomed if individuals block our view, even if the ultimate aim of analysis and action is a betterment of their welfare. The general point also underscores ecological analyses that emphasize the determining impact of environmental variables (such as the constraints imposed by carrying capacities) on human behaviour and the structuring of societies and economies.

These viewpoints interact with the internal currents of many of the disciplinary discourses relevant to understanding environmental problems. Traditional historians focused on individuals, usually political leaders, as in a different fashion have sociologically inspired historians more recently in their studies of ordinary folk. Without this constant concern for individuals, Turner has argued, a sense of human agency disappears, and history becomes merely a product of the interplay of impersonal forces (1999: 303).

Individualistic approaches have been prominent too in the history of sociology. However, others focusing on norms, roles and rules, and the structures that limit an individual's capacity to shape his or her environment, have probably been a more dominant feature of the discipline (Holton, 1996: 35; Healy, 1998). The latter draw on Durkheim's emphasis on 'social facts' and on the 'external constraints' these impose on individuals. In counter-traditions can be seen the continuing influence of Weber's individualism, or of the ethnomethodological (and determinedly anti-theoretical) strategy of observing people striving to cope with daily living, and continually questioning and creating social relations as they do so. Giddens's attempt in the 1980s in structuration theory to effect a fresh resolution of the agency-structure problem reflects these and other influences.

Tensions on this issue have been less marked within economics, especially since its disciplinary shift away from the broader frameworks of political economy. Two separable metasets of assumptions are important: first, that individuals (or other actors in micro-analyses such as firms) are rational beings with preferences, who calculate utilities and make decisions on these bases; and, second, that what they do has a bearing – beneficial, provided interfering monitors and cliques such as big governments and labour unions stay out of the game – on the larger social and economic worlds in which they live. If we want to eat, as Smith puts it, we rely on the self-interest, not the community-spirited altruism, of the butcher, brewer and baker. For good things to happen in a society, according to Hayek, social agreement on common objectives is thus not necessary, and attempts to bring it about are likely to be counterproductive. These kinds of arguments have tended to deter many environmentalist observers from treating rational-choice perspectives seriously. They may also, depending on the definitions of key terms, add up to no more than tautologies, especially since any thought or action can with ingenuity be construed as the product of self-interest.

The individualist issue has left its mark on IR theory, though again in different ways. Reinterpreted notions of rational choice have traditionally been central to realist and neorealist approaches to understanding the behaviour of states and other actors, in particular through study of the decisions of political leaders. A critic from sociology has argued that the field of IR persisted with an 'individualist attitude' long after such thinking fell into disrepute elsewhere in the social sciences, including study of the domestic politics of states (Ginsberg, 1974: 167). The tendency perhaps reflected early influences on IR, particularly from diplomatic history. It did not long survive the rise of the realist paradigm in the 1940s. 'Levels' of analysis, including that of the state, took on distinctive attributes, and were held to be associated with differing explanatory variables, normative values and methodological approaches. Analysts took care to avoid confusing individual transactions with those of states, whether for normative or empirical purposes. The structures within which individuals acted were assumed to be more promising foci of research. Despite recurrent charges of reductionism or a spurious foundationalism, traces of individualism have nonetheless remained. For example, there has been continuing scholarly interest in the idiosyncratic influences on foreign policies of specific individuals. However, this has tended to mean study of political leaders, a Bismarck or a Kissinger, or the interactions of foreign-ministry bureaucrats. The approach does not extend, say, to the crop decisions of a Bolivian farmer. The re-emergence of normative enquiry in IR, and responses to some strands of critical theory (though not, at least not directly, the variety that decentres individuals), are among developments that have countered this tendency. Concepts of security have been increasingly restructured and 're-levelled' to focus not on the attributes and supposed needs of states, but instead on the health, employment, economic or environmental needs of individuals.

Environmentalist arguments have often been attracted to individualist formulations. These in turn tend to lead to investigation of the structures that both enable and constrain actions. A farmer in sub-Saharan Africa may contribute to desertification by exhausting firewood supplies, just as one in Europe may foster biodiversity loss by destroying hedgerows. Neither set of actions can be understood without accounting for the economic structures, from local to regional and global, within which individuals make decisions, and the broader constitutional, legal and political frameworks of societies. Individuals, moreover, are multifaceted. One in a large European city may recycle wastes, but that person may also engage in a wide variety of energy-consumption, food-purchase and other activities that, aggregated across those of all other residents, have significant consequences for the size and character of the city's ecological footprint. An individual's expectations may also be at odds with the actual outcomes of actions, especially in view of the complex rings of consequences associated with individual actions in a global economy.

A notion of individual agency remains nonetheless central to much writing on environmental issues. Indeed: 'In any overarching theory of environmentalism', Spybey argues, 'the fundamental principle must be that the individual stands in some kind of relationship with the environment' (1996: 149). Pictures of this relationship vary. Images of individuals differ among disciplines. Individuals exert a wide variety of direct and mediated effects on their environments. Much social theorizing on environmental questions has accordingly tended to pursue an indirect strategy. More general questions of the relations among individuals and social actors are taken as the primary ones for enquiry, and it is assumed that this is also the route to understanding and ameliorating the environmental effects of human activities. I will take these themes further through discussions of the applicability to environmental problems of, first, rational-choice approaches and, second, approaches from the sociological study of individual agency.

Choice environments

In a sense the claim that a framework of rational expectations and choice can usefully be applied to the study of environmental questions follows by definition from its own assumptions. Rational choice 'has no thematic limitations. ... It provides a coherent methodology which allows researchers in different fields to have a common purpose and common research agenda to facilitate dialogue across diverse fields' (Dowding and King, 1995: 16). From their origins in microeconomic analyses of choices in markets by consumers and producers – individuals or firms – and their consequences for markets and other matters, approaches based on choice and expectations have been extended beyond that discipline's boundaries in a combination of push by economists and pull by researchers in other disciplines. This expansion process sometimes makes it difficult to distinguish transdisciplinary insights from those associated more specifically with the cultures of economics. Illuminated, nonetheless, through such means have been problems as diverse as educational achievement, religious affiliation, the choice of sexual partners, and criminal behaviour, and the behaviours of actors ranging from voters and lovers to superpowers and multinational oil companies. From this perspective, that is, environmental questions present no special problems of applicability. Local difficulties are manageable by making adjustments with tools such as the concept of bounded rationality. A subjective-rationality approach, for example, can treat behaviour as rational even if the beliefs on which individuals base actions are false.

How far, if at all, this is a useful guide for understanding individuals remains an issue that divides observers and disciplines. Key words such as 'interest', 'rationality' and 'choice' are in practice highly contested. It is not self-evident, for example, where preference-sets come from. If these can be shown to be structured in practice by large forces external to individuals – firms, governments, discourses – then a rational-choice framework may no longer be a rational choice for the observer. To the extent that environmental reasoning rests on assumptions about cooperation, community, trust and interpersonal communication, and the restraint of self-interest in the production and consumption of goods, rational-choice frameworks appear to be limited or just wrong. The stunted and abstract individuals in them appear different from those in environmental discourses. Choice frameworks can nonetheless in principle capture most of the kinds of actions that environmentalist critics might want to protect from their over-zealous reach. These include study of the interest a rational environmentalist might have in not buying over-packaged products, or in blocking a logging company's trucks' access to a piece of old-growth forest. We can look more specifically at examples of the study of environmentally significant individual choice in relation to four areas: public policy, markets, social structures and ecosystems.

In the first of these, approaches to public policy, individuals have preferences about the environmental (and other) policies of governments, and, by extension, those of groupings of states in IGOs. In principle it is possible to draw up a preference function for each individual which would describe what people want policies to be and indicate their rankings among these. Such functions can accommodate the fact that individuals vary, perhaps considerably, in the importance they attach to different policy areas. Environmental policies count a lot for environmentalists. Other policy areas may matter much less, or come into an environmentalist citizen's accounting through more circuitous routes. Sparing education budgets from the rigours of government deficit-cutting measures may mean even less money is available for environmental protection. Certain expenditures on defence or agriculture may be argued to be causally connected with actions deleterious to the environment. In general, we could expect that a person will respond disproportionately more to small changes of government policy in relation to an issue ranked highly in importance (Hinich and Munger, 1997: 26, 55).

Individuals are thus linked to public policies in complex webs of expectations. Government policies create mixes of incentives for agents, including individuals and firms, who tend to respond to new programmes and policies by acting in ways that they think will increase the benefit, or decrease the headaches, they expect to get from them. Coate and Morris argue that these kinds of mechanisms are the key to understanding why government policies persist (1999: 1327–8). In some countries, compliance with or wider political support for policies is associated with tax earmarking, or the setting aside of revenues from environmental taxes for specific water-quality, clean-up and other sustainability projects (Marsiliani and Renström, 2000: C123–4). A recent study of Mexico suggests that whether or not companies comply with environmental regulations depends on many factors, including disincentives created by the absence or sporadic character of monitoring and enforcement by government inspectors, and incentives resulting from both environmental regimes and extra-legal factors. The crucial element, though, is the attentiveness to environmental policy questions on the part of individual managers that results from training in these areas (Dasgupta et al., 2000).

This approach encounters at least three kinds of difficulties: describing preference functions accurately, modelling the translation of preferences into policies, and gauging the effects of government policies.

On the first of these, preferences may be expressed obliquely as stands for or against political parties. Responses to the Greens in the 1998 German elections, for example, were shaped in part by the party's platform plank on increased fuel taxes. General expressions of diffuse support for environmental-conservation goals may shift when a government's (or party's) declaratory statements are given detailed expression. These may involve the creation of restricted-access protected habitats, additional land-use rules, more stringent quota arrangements for coastal fisheries, or tighter energy-use regulations. Policies may thus affect, reveal, or conflict with, other preferences of individuals. An individual's support for protecting a wildlife species might implicitly be dependent on her being able to visit a protected area and see the results of conservation measures; it may evaporate if in practice conservation projects bar such access. Conversely, soft objections to environmentalism can change if a preferred relaxation of government regulations leads to a noticeable increase in health problems brought about by air pollution, or in threats to fisheries from river-borne effluents.

While this problem could arguably be resolved by the use of more sensitive and more continuous probings of public attitudes, the second is problematic in a more fundamental way. In social-choice terms, people's preferences cannot be translated directly into a transitive aggregative list that is some kind of mirror of the range and character of individual choices. However, this is less a predicament for environmentalism than it is for democratic theory generally. A democratic theorist, or constitution- or electoral-system designer, has to grapple with the facts of diverse and conflicting individual preferences on across-the-board issues and parties, and compare the match that public policies or election results achieve with these. Environmental advocates may be democrats, but they want to get certain things done. In environmental theory, preferences that suggest anti-environmentalist bias – private-sector resistance to or reinterpretation of scientific findings on the toxicity of a product, for example, or an industrial association's criticisms of regulatory environmentalist options for potentially infringing on intellectual property rights – have somehow to be refashioned (by regulation, round-table compromise, education or other instruments) rather than just accepted as givens in grand utilitarian or social-discourse projects aimed at eliciting the democratic choices of societies.

Both perspectives have a bearing, thirdly, on the effects of government policies, including assessments of the consequences of international environmental regimes. The national implementation of multilaterally negotiated agreements, for example on the ozone layer or climate change, rests initially on government legislative or regulatory frameworks, but then depends on the choices made by individuals and firms. Indeed, one implication of choice approaches derived from the study of economic agents is that government policies are in a sense bound to be ineffective, because players adjust their actions according to rational expectations of those of governments and other players.

This brings us to a second area of individual environmental choice, in relation to markets. The institution remains a contentious one in environmental discourses. Their defenders argue that markets work more effectively towards environmental-protection goals than government interventions in command-and-control systems (Beder, 1996: 51–2). Markets fail, that is, not because of some inherent flaw in the institution, but because of their uneven spread or the lack of uniform application of property concepts, for example in the inadequate or non-existent pricing of natural resources.

At the individual level, this raises the question of the extent to which ecologically rational consumer choice is feasible (Wagner, 1997). Some forms of market response by individuals contribute to problem-solving of the kind valued in traditional environmental discourse. On the supply side, for example, individuals can set up small organic-food or bicycle-repair businesses. Individual choices have consequences for future generations of consumers, a constituency sustainable-development advocates claim to be able to represent. Treating these as economic agents, however, and developing methods for expressing their demand for goods, remains fraught with difficulties (Martinez-Alier, 1987: 158–9). Some aspects of the distant-future consequences of current choices, though, for example in relation to carbon-dioxide build-up in the atmosphere, can in principle be quantified and could thus enter into energy-related individual choices (Azar and Holmberg, 1995). More generally, imperfect information tends in practice to limit the application of rational-expectations approaches to these problems.

Thirdly, individual environmental choice operates in relation to societal structures and processes, and thus indirectly impacts on resources and the natural environments of economies. In addition to having preferences about government policies, that is, individuals make choices among civil-society actors, including environmental NGOs. They distribute elements of support for such organizations – the intensity of their commitment, donations of time, financial contributions, and decisions to lobby or engage in direct-action or civil-disobedience campaigns – and make adjustments to these decisions over time. Rational choices are often linked in such cases with expectations of reciprocal responses by others. Chains of such exchanges lead to the formation of groups and communities. There is a related market for the ideas of environmentalism. Rational choice has been used as a means of studying individual decisions about support for different religious beliefs (Iannaccone, 1997: 26–7). Similar processes are arguably at work in relation to the environmental-ideas market. These affect decisions by political parties on the inclusion of environmental policy planks in election platforms.

Finally, there are apparently direct relations between individuals and environments, unmediated by social groups, market forces or public policies. Even the more obvious of these, however – household waste-treatment practices, vegetable gardening, admiration of landscape, energy use from locally available supplies of firewood, and so on – remain tied to the larger economic and societal, and global, contexts of individual action.

Extensions of rational choice to environmental problems thus have the much-proclaimed advantages of the original imagery of economics. Theoretical elaborations are parsimonious; applications imply a notion of individual dignity by emphasizing the power that comes from choice; they blend creatively with the argument that good things in society do not have to be planned, and preferably should not be; and they facilitate predictions about the effectiveness of government policies, including policies in relation to international environmental regimes, when these impact on the choices made by individuals. However, choice metaphors are traditionally contentious within environmental debates. Their presence is often more indicative of major fault-lines than of consensus on the requirements of good environmental enquiry and action.

Firstly, the individuals who live in choice worlds tend to be creatures alien to those in environmentalists' habitats. Their apparent abstraction and isolation – beings 'set off from others ... "over and against" the world' (McMahon, 1997: 164, 166) – tends to repel environmentalists and other critics who prefer their individuals real, live, community-dwelling and multidimensional. More specifically, the pursuit of 'interest' by individuals in choice accounts blurs with the negative connotations of terms like self-interest and selfishness, especially when these are juxtaposed with the lexicons of calculation, strategy and games. This unease rests partly on misunderstandings, however, which in turn reflect misgivings about the traditional practices of the discipline of economics. Thus an individual can be other-regarding when making rational, self-interested decisions (Abell, 1996: 265). Players can opt for cooperation. There is no reason from a theoretical perspective why a choice approach could not accommodate a wide range of environmental goods – spiritual enrichment from wilderness, for example, improvements to air quality in large cities of the South, or care for the well-being of future generations – when looking at an individual's preference function and rational decision-making.

Secondly, the dependence of choice on information presents often fundamental problems for environmentalist critics. Individuals typically lack, or may have little interest in obtaining, data on the complex chains of connections linking their decisions and the options facing them with structures and processes in diverse sectors of the global economy. The consequences or sources of certain kinds of choices may thus remain hidden. A preference by a consumer for pleasingly and uniformly shaped and coloured carrots or tomatoes could be taken as implying rational support for the use by agricultural producers of the pesticides and other practices required to generate these appearances. In practice this is unlikely to be a dominant feature of most individual consumer decisions. Environmental NGOs accordingly try to focus on such instants of choice, for example through campaigns linking North American fast-food consumption habits with Central American forest destruction, or by advocating improved labelling to include reference to GM products. However, other contexts, particularly tobacco, suggest that risk communication by these methods may not guarantee wise individual choice because of the multiple messages impinging on decisions from social groups and other sources (Leiss, 2001: ch. 7).

The problem is more complex if in addition to data the information requirement is interpreted to include scientific knowledge. Assessments of the probable consequences of individual actions would then call for judgements on such questions as the climate-change impacts of different packages of household energy use, the implications for ecosystems of different industrial or agricultural processes, and the health consequences of pesticide spray drift from lawns and golf courses or of potential gene drift from the cultivation of genetically modified plants. Even if individuals wanted and could digest such knowledge, the nature of scientific decision-making makes an approach to certainty on the risk involved in such issues unlikely. Further, such knowledge may in any case be unattainable even in principle because of the inherent indeterminacies of economic, social and ecological systems (Faucheux and Froger, 1995: 33). On the other hand, the fact that information is necessarily limited is not an argument against societies, and individuals, taking steps to make the information at their disposal less imperfect.

This, thirdly, brings us to the subjective variables that enter into individual decision-making and post-decisional evaluations. When Y2K problems failed to materialize in January 2000, a widespread response was that this was evidence that the problem had not in fact existed, and that the investment of resources in trying to solve it had therefore been wasted. Up to a point, as with other objections to choice theory, the approach can handle subjective problems. For example, it can be adjusted to incorporate the subjective factors that typically enter into voting decisions. Variables in environmental debates such as wishful thinking, panic or superstition, can also be accommodated by rationality frameworks. At some point, though, the process of building in second and other thoughts begins to smudge the clarity and parsimony of rational-choice insights. Especially when institutional complexities enter the picture, the exercise takes on a more decidedly sociological, as opposed to economic, complexion. A further adaptation, then, is to recognize that individuals make environmentally significant decisions in part on the basis of habit or inertia or in response to social pressures, and to shift to a bounded conception of rationality. These factors can then be studied by means of interdependent decision models, or by isolating – and redefining as part of the rational calculus – those parts of choices that flow from observation and imitation of the behaviour of others (Abell, 1996: 266, 268).

A fourth set of reservations focuses on the sources of preferences and perceived constraints. In many circumstances the limitations on an individual's capacity for choice are such as to invalidate the use of 'choice' language to describe what happens. Individuals vary in their capacity to overcome, even if they recognize, the constraints on their decision-making, for example by shifting the costs of decisions on to other agents or imposing limits on the decisional capabilities and resources of others. Many decisions are made in the contexts of groups of one kind or another. The processes by which individuals in small groups arrive at decisions have similarities, whether the group in question is a sports club or a committee of foreign-policy crisis decision-makers (Nicholson, 1995: 164). Individuals also play multiple roles. Environmental preference functions vary, accordingly, with the different contexts of action in which individuals find themselves. A Greenpeace supporter may have a job in a chemical plant, and a scientific expert may change views as well as hats when consulted by different national governments and international agencies.

A final set of issues relates to the macro-consequences of micro-decisions. There is ongoing debate in economics on the extent to which study of aggregative microeconomic activities can adequately explain the macro-properties of economies. Even if the tyranny of small decisions is assumed, outcomes typically differ from the anticipated consequences of actions. Choices have consequences for ecosystems, but these are not necessarily those intended by an individual (or the environmental group whose actions she supports). And structures shape outcomes. Preferred environmental consequences are unlikely to be realizable for an individual living under a military dictatorship or conditions of extreme poverty, or awaiting destruction by an asteroid. Many environmentalist arguments nonetheless tend to share with choice theory the view that micro-decisions can indeed be aggregated across societies and economies. Ecological-economic modelling rests at some point on the assumption that both the short- and long-term characteristics of economies and societies are somehow 'structured by the decisions of individuals' (Folke and Kåberger, 1991: 274). Individuals may see their own actions in this light, a disposition encouraged by environmental NGOs. An individual can opt to join or send a financial contribution to Greenpeace in the expectation that it will become 'my "hired gun" in the fight against the whalers' (Gibbins, quoted in Wilson, 1992: 112).

However, at some point this process becomes constrained by distance effects. The gulf between individuals on the one hand, and global-level developments on the other, limits the usefulness of choice models that are either restricted implicitly to local or national societies, or which do not specify connecting institutional variables associated with states, companies or transnational civil-society groups in choice webs. Environmentalist critiques also diverge from rational-choice accounts in locating the sources of or incentives to ecologically benign actions. Many question reliance on market mechanisms, or the encouragement of industry self-regulation. Choice approaches from this perspective are weak not necessarily because of any inherent methodological flaws, but rather because of their association with theoretical frameworks derived from the cultures of economic reasoning that respect markets as the primary allocative mechanism for societies. The beliefs about economic progress that arguably sustain these (Nelson, 1997: 187–8) are likewise found wanting.

Finally, these considerations raise again the issue of the relations between actual and anticipated outcomes. As Jervis notes in his discussion of international 'system effects', these are often divergent. He cites the example of the Reagan administration's intention to reduce abortions in the South by cutting development-assistance funding to clinics; these, however, also sold condoms, so the outcome was an increase in abortions (Jervis, 1997: 61–7). The environmental actions of individuals and organizations can also be linked to paradoxical consequences. However, these are often more apparent than real. In practice many dissolve into differing interpretations of either the scope and coherence of environmental goals, or of the nature of the causal connections among key sets of variables. Examples of these kinds of effects in the late 1990s were the threats to voles in England resulting from increased populations of mink following releases from fur farms by animal-rights groups, and the hazards to eagles and other wild birds caused by the expanded use of windmills for alternative energy in California and Scotland. Pursuit of one environmental goal, that is, affects the achievement of others. However, there is no consensus on what such a package of 'environmental' goals should contain; and even if, hypothetically, such a consensus could be achieved, their pursuit would still require political decisions among conflicting options. So surprising and paradoxical outcomes, and apparently self-defeating actions, would presumably persist.

Debates on environmental policies often contain critiques of the allegedly self-defeating character of non-preferred options. Thus from a market or neoliberal perspective, pursuit of some environmentalist strategies (for example, anti-consumerist stances generally, excessive regulation of industry and agriculture, environmentally based non-tariff barriers, North–South sharing of environmental technologies, or inflexible barriers to mining or other resource activities in protected areas) is argued to lead to the marginal economic impoverishment that hinders environmental policy progress. Sustainable-development perspectives tend to maintain, among other things, that the lack of entrenchment of sound environmental thinking in the private sector leads eventually to lower profits, social and political instabilities injurious to economic activities, and long-term checks to economic growth.

The environmentalist observer is thus often in a different situation from that of the choice analyst. The rational-choice approach, in a sense, is that of the democratic, and more specifically the utilitarian, theorist. The substantive content of particular preference functions and decisions is not the concern. For the environmental analyst, the situation is reversed. The substance of preferences and decisions is the primary issue. The environmentalist wants to retain a capacity to influence the formation of these by individuals.

The quest for agency

The individual who emerged in the Western philosophical tradition in the seventeenth and eighteenth centuries has led a fretful existence in sociological thought. For sociologists, a realist appreciation of the social and economic forces that buffet individuals has often competed for the centre stage of enquiry with a stubborn reluctance to let go of the lives of (ordinary) people, and, more specifically, of the conviction that what they do and think matters and shapes social worlds. Notions of agency have been moulded by exposure to images of the power of economic and social structures. In formulations influenced by the classical Durkheimian tradition, these exist in a world external to individuals that coaxes, manipulates and coerces them. This world pushes them into being and action, in part by blocking off some possibilities. A human being, as Marx wrote, is 'not merely a gregarious animal, but an animal which can individuate itself only in the midst of society' (1973: 84).

Agency-oriented attempts have tried to see structures not as givens, but as somehow the products of individual actions and interactions, or as the results of collective societal responses to the need for rules and institutions brought about by these actions. However, so seductive has been the study of big frameworks in some areas of sociology that writers have at times neglected to connect these persuasively with explanations of individual actions and understanding of their effects. Awareness of the interconnectedness between agency and structure, Emirbayer and Mische argue (1998: 963), has led to a failure by many sociological theorists to develop theoretical perspectives that focus specifically on agency as an analytical category. Some sociologists have responded by investigating ways of reconciling elements of rational choice with key features of theoretical sociology (Abell, 1996; Goldthorpe, 1998; Zafirovski, 1999). There have been complementary attempts from inside economics. Individuals, Avio argues, are social and communicating beings as well as economic agents. Thus:

> it would be incorrect to assume that social agents *choose* between two mutually exclusive states of being: *homo oeconomicus* or *homo communicandus*. Socializing processes and the pragmatic nature of language assure that we have communicative competences, including the know-how needed to reach a mutual understanding. Since *homo oeconomicus* is overlaid upon *homo communicandus*, the former cannot avoid *also* being the latter.
>
> (Avio, 1999: 522–3)

What does this turbulent record have to tell us about individual agency, understood outside the choice frameworks of the economist, in relation to natural environments? In approaching this question I will focus particularly on the structurationist tradition of enquiry initiated by Giddens's writings of the 1980s.

Giddens's work on structuration represents a return to older notions of agency, particularly where he exults in a robust individualism. He repeatedly draws our attention to the 'knowledgeable human agent', a person neglected, he argues, in Parsons's approach to the understanding of action (1984: xvi, xxxvii). Agency, he emphasizes, is a concept restricted to the level of individuals. Collective agents, by definition, cannot exist. We can give descriptions of the actions of groups or collectivities, that is, but 'only individuals, beings which have a corporeal existence, are agents' (1984: 220). Both the diverse theoretical perspectives that rely on structuralist accounts, linguistic and economic, and those in post-structuralist and other narratives that effect a dissolve of individuals, thus share a common failure to focus on the central image of the individual agent (Giddens, 1987: 195–6).

These actors are knowledgeable, then, or, in a favourite Giddens word, competent. Unlike those in class analysis whose lives tend to be stunted by false consciousness, these individuals understand the world, themselves and their actions. They 'routinely and for the most part without fuss ... maintain a continuing "theoretical understanding" of the grounds of their activity' (Giddens, 1984: 6). As Thompson summarizes the image, individuals are knowledgeable agents 'who are capable of accounting for their action: they are neither "cultural dopes" nor mere "supports" of social relations, but are skilful actors who know a great deal about the world in which they act' (1989: 58).

This sense of the individual as a comprehender of the world echoes Weber, and is crucial to Giddens's depiction of the relation between agency and structure. The latter, in a sense, collapses into a relationship activated by the former. Despite conflicting traditions in sociological thought that have respectively stressed one or the other, he argues that the emphases are not mutually incompatible. Structural explanations on the one hand, and methodological individualism on the other, are not necessarily conflicting theoretical approaches (Giddens, 1984: 220). Instead, their respective foci become in a sense fused. Dualism dissolves into duality. Individuals through their actions play a part in the creation of structures (or what he calls the 'structural properties of social systems'). Such features do not shape or in some fashion determine individual actions; rather they are the 'medium' of the actions that constitute the features of societies. Structure has a dual character. The structural properties of social systems 'are both medium and outcome of the practices they recursively organize. Structure is not "external" to individuals: as memory traces, and as instantiated in social practices, it is in a certain sense more "internal" than exterior to their activities in a Durkheimian sense.' Further, and in contrast to structuralist accounts that stress the limitations on individual action, structure 'is not to be equated with constraint but is always both constraining and enabling' (1984: 25). Structures thus assist individuals in performing the acts that constitute them.

How much further does this strategy take us in relation to the environment? Perhaps not very far. In terms of its direct usefulness on this point, three limitations arise. Firstly, as for other sociologists in the 1980s, and those since who have worked to develop structurationist perspectives, the environment is not a central question in Giddens's work. While he does deal with issues of the environmental costs of late modernity, this concern tends to be less prominent in his handling of the problems of (individual) agency than it is in his critiques of modernity more broadly. Perhaps since modernity is viewed as inherently globalizing, his accounts of it tend to be skewed in grander directions, though he retains an analytical sympathy with individuals and the problems of psychological stress and intimacy they experience under its conditions. In relation to the environment, Giddens's approach at times takes on something also of the nineteenth-century romantic texture of 'modern Man bereft of the consolation of Nature'. This limits the potential for elaboration of a model of the ways competent individuals interact with ecological processes. Indeed, in a sense even the possibility of this kind of interaction disappears from his analysis of modernity. An understanding of ecological processes becomes in a sense superfluous. In the transformation of nature by modern economies and societies, that is, the natural world becomes part of the created environment. It thus consists of 'humanly structured systems whose motive power and dynamics derive from socially organized knowledge-claims rather than from influences exogenous to human activity' (1991: 144). This perspective leads to some intriguing dissonances within the overall structurationist framework. Since natural environmental systems are part of the built, or at least the manipulated environment of societies, we presumably cannot look to structuration theory for guidance about the ecological consequences of human actions. In a sense these disappear into the traditionalist worlds of classical social anthropology.

Environmental groups, moreover, occupy in Giddens's accounts of structuration theory and modernity pivotal positions as elements of one of a number of countermovements in capitalist societies. This is a perspective that tends to ignore the complexities of environmentalism, for example the ways in which various forms have become embedded in state structures. It also in a sense reduces environmentalists to the performance of an interchangeable critical or checking role alongside other social movements. The multiple and conflicting representations by individuals who are members of such groups – of the nature of the environmental problems they are addressing, and the likely outcomes of group activities – do not conveniently fit Giddens's own neater conception of them and of their world.

A second constraint is that Giddens's account of structuration theory, regardless of how it touches on environmental topics, is not necessarily designed as a useful bag of tools for empirical researchers. Giddens's cautionary tone on the question of applicability follows logically from the heuristic character of the arguments. The argument that social structures, however defined, are constituted by individual agency through the medium of structural laws is not in itself an empirically testable proposition. Nor is it clear how it could lend itself to the production of these. The eclecticism of the theory also bears the imprints of multiple theoretical influences, which in practice, if not in principle, makes application problematic. Theorizing, in a sense, becomes for Giddens a different sort of activity from empirical research, and one that should not be judged on the criterion of its capacity to make contributions to that kind of enquiry. Structuration theory should be used selectively, that is, and seen 'more as a sensitizing device' than a self-help empirical research manual (Giddens, 1989: 253, 294). Other writers have indeed used his work in a variety of empirical and theoretical ways to explore environmental topics. Goldblatt, though, focuses on Giddens's historical analyses of capitalism, the nation state and globalization. He develops an environmental critique based in part on the need for better understanding of ecological processes, a feature that several critics have argued is conspicuously absent in Giddens. However, he does not orient this around the core concepts of individual agency and structuration that are arguably the foundational elements of Giddens's writings on structure (Goldblatt, 1996: chs 1 and 2).

A third set of problems relates to the knowledge of agents. The point is crucial for Giddens's identification of the processes involved in the structuration of social properties. It also serves to differentiate individuals from social organizations, and is central to the attribution of agency exclusively to the former. Two considerations, however, limit the usefulness of the argument. These derive respectively from the diverse and complex character of, firstly, ecological processes, including interactions among economic, environmental, demographic, cultural and other variables; and, secondly, of the global processes, including economic globalization, that impact on (or, in structurationist accounts, arguably flow in part from) the actions of individual agents. The structurationist conception of agency is nonetheless a helpful route into understanding the relations of individuals in circumstances of globality.

As figures in globalization processes, individuals have tended to present themselves more in sociological accounts than in those of other social sciences. Some features of globalization can be seen in this context as analogous to traditional sociological treatments of structural variables. Thus globality is commonly portrayed in a Durkheimian social-fact manner that highlights the external forces that constrain individuals, cultures and the activities of groups in civil society. Alternatively, globality is also interpreted as being integrally connected with individuation and agency. As Spybey puts it: 'The single most important factor in globalization is that the global enters into our local day-to-day reproductions of social institutions.' As a result of the globalization of political, economic and cultural institutions, 'there is virtually no one on the planet who can participate in social life without reference to globalized institutions in some form or other' (Spybey, 1996: 9, 151). Dombrowski similarly views individual agency as central to globalization, and, like Giddens, and despite the systems' complexities of global society, sees individuals as the only economic or social agents. 'Only individuals choose to withhold or transfer allegiances within accepted political processes, to vote with their feet by migrating, or to rebel violently. Only individuals can devise organizational and institutional forms (from states to corporations to markets) necessary to achieve personal and social goals' (1998: 388).

The issue is thus whether, and how, individual agency has a bearing on global environmental and other processes, as opposed to the structural features of the more immediate domains of individuals. A constitutive notion of individual agency can be retained by focusing on the way the predicaments and choices of individuals continually *interact with*, as opposed to merely being *conditioned by*, global economic, social and cultural forces. Giddens identifies the risks that confront individuals as a result of 'burgeoning processes of globalization'. These range from chemicals in foods to nuclear accidents. They are part of the 'climate of risk' in which individuals live in late modernity and the 'crisis-prone nature' of this historical condition (1991: 122–3, 184). However, the complex chains of cause and effect associated with globalization are not easy to disentangle (Dombrowski, 1998: 376). Some of these analytical difficulties echo classical problems of methodological individualism: If both individual acts and social properties can be observed, how can the proposition that the former cause the latter be tested empirically? What are the grounds for rejecting the view that structures govern habits, which condition individual choice? Is agency a property that belongs to more powerful individuals, in which case it collapses into tautologies or concepts of power?

In relation to global as opposed to neighbourhood or other 'closer' settings, moreover, the possibility of designing realistic models of the collapsing dualisms of agency and structure seems more remote. The social and political distance of individuals from structures reinforces the analytical attractiveness of more externalist conceptions of structure (Healy, 1998: 512). Even within national societies, some observations on the environmental consequences of individual actions look true primarily by virtue of the imaginative or tacit use of thought experiments: *if* all (or a significant number of) individuals were to change their activities (watching movies, voting, eating tuna) *then* larger-scale effects on societies and economies would follow. The tautological element here derives from the conflating of simple aggregation with causality. However, this is not necessarily true of all formulations. Structurationist ecological theory can accommodate factors such as those contributing to the internalization of changed habits, the social spread of these through mimicry and other processes, spillover effects between changes in discrete sets of actions by individuals, and the multiple flows of effects between actions and agents' sources of information and ideas.

A more sustainable theoretical approach, however, requires a partial accommodation with some of the key elements of rational expectations discussed earlier in this chapter. For example, individuals, as owners of multiple affiliation options, increasingly make identity choices in a globalizing world in ways amenable to study by way of both choice and sociological-agency approaches. These options include the choice of active affiliation and identification with transnational environmental NGOs. Interpretations emphasizing actors' expectations of reciprocity, or, more generally, the systemic consequences of micro-level choices, likewise straddle both sets of approaches. Globalization complicates these links. Spybey writes that 'the greater the global whole, the more opportunities there are for the individual' (1996: 169). Dombrowski is likewise alert to the power of global structures, but sees these too as partially constituted at micro-levels. Individualization is 'important for understanding global transformations, and international relations more generally, because its historical development has allowed larger numbers of people in a wider variety of societies to make, or at least try to make, meaningful choices about the types of polities to join, support or even die for' (1998: 364). However, the range and character of opportunities for individuals to demonstrate efficacy is limited. Crucially, it may exclude influence over the rules governing the workings of globalizing economies and the institutions fostering these processes.

Individualism and global ecology

Methodological individualism has occupied an unstable ground hemmed in by economic, linguistic and other structuralisms on the one hand, and subjectivist decentring post-structuralisms on the other. Yet it has maintained a lively capacity to inform and periodically reinvigorate analyses. It remains central to many regions of environmental discourse. The social-action and policy-advocacy function of environmental theory, however, as well as concerns about the effects of structures on the shaping of individuals' preferences, make for some incongruities between environmentalist and rational-choice theory. Attempts to reconcile elements of traditional sociological and economic reasoning on these questions are thus important routes for the development of environmentalist argument and practice.

The global level of analysis limits the applicability of such approaches because of its distance and externality from individuals (except, perhaps, the individual considered in the traditional IR sense of a political leader or diplomat acting for a state). Activities at this level more obviously shape the lives of individuals, who are affected by international trade flows and decisions and MNC investment strategies, than they reflect micro-level actions. Notions of constituting processes derived from sociological frameworks are nonetheless applicable at this level through extensions of the causal chains connecting individual agency with structural variables. Aggregated individual decisions can thus be seen to lead to emergent properties and other macro-level consequences. Useful in this task are geographical or spatial notions of globality, which help focus attention on micro-levels of action and their interactions with global frameworks.

7
Brief Authority

Collectivities of various kinds are part of the problematic of individual agency discussed in the previous chapter. As the power of economic and political structures is seen to grow, the capacity of persons to induce microgenic change is diminished. For radical social change to occur, strategies of collective as opposed to individual action are accordingly required. By contrast, the focus in this chapter is on the actions and interactions of larger-scale collective actors – NGOs, states, corporations, governments and their bureaucratic component parts, and intergovernmental agencies. While the activities of these have implications for individuals, and may on the face of it be designed either to promote or manipulate their interests, these questions are not the concern here. These organizations have complex internal features. They interact in diverse sustained and temporary ways, and they do so increasingly in a global or transnational manner. Insights from sociology and social theory on the one side, and from political economy on the other, are thus particularly useful for studying the complex threads of governance. Different actors in global society have governance claims relevant to environmental problem-solving. After discussing these, the chapter examines in particular functionalist ideas of governance as guides for the environmental domain.

Conundrums of governance

The games these actors play are oriented partly around the making, interpreting, avoidance and change of the tacit and explicit rules that ostensibly govern their interactions. The upshot is not government in the sense in which this would be recognized inside national societies. International law has features that distinguish it from law considered as part of the domestic order of polities. Governance considered as a process, however, or as interconnected sequences of games played or contracts negotiated by multiple actors, is nonetheless increasingly a structural feature of contemporary global society.

From an environmental perspective this development has both beneficial and less desirable consequences. Many multilateral forums and organizations now deal with environmental issues. This constitutes a significant setting for environmental policy-making and societal action inside states. The external context both constrains and promotes domestic activities. The growth of environmental governance functions at interstate levels is itself also part of a larger set of changes in global society. Particularly in the context of globalization, these changes extend the reach of economic actors and their capacity to influence the actions of national governments. The process encourages the further retreat of the state from environmental and social-welfare-related functions, in part by both facilitating and responding to global discourses that prioritize conventional economic-growth and trade-liberalization goals. These developments do not necessarily undermine the pursuit of environmental policy goals. Some conspicuous forms of green objectives, such as measures against air pollution, have been built into the contemporary liberal state. Alternately provoking action by, and encouraging state collaboration with, civil-society actors, globalization has multiple ecological consequences. Systemic changes have nonetheless transformed the contexts in which environmental decisions are made by NGOs and governments.

Issues of governance have long been part of IR, and an important subsidiary discourse in IPE. The sociological study of globalization has also increasingly turned to these themes. A perennially core question has been the nature of the primary units of macrosystems, and the factors that structure relations among actors. In classical IR, the main constituting role is performed by states. These possess territory, natural resources, national allegiances and weapons. Their repertoires include war. This institution shapes the relations of states, patterns of alliance-making, changes in territorial boundaries, the creation and extinction of states, and features of their internal regimes. The standard account also guides observations across time. Patterns of rise and decline among powerful states highlight significant mechanisms for understanding evolutionary change in the international system and the 'long cycles' of global politics.

Thus in IR-realist perpectives, IGOs and international law are more the creatures of states than influences on them. Peripherally and temporarily useful devices, they are routinely blanched of the authority attributes that some reform-minded states periodically try to inject into them. The capacities of civil-society actors are also restricted in these perspectives. In normative terms, they are constrained by the presumption that the state is coterminous with the political community and is its representative in dealings with outsiders. This genre of interpretation has traditionally demoted environmental, social and cultural matters, or periodically resuscitated them through expanded neorealist conceptions of state interests. Cooperative arrangements between states, including IGO activities, are thus best understood as products of the interplay of the pursuit by states of their respective interests. Outcomes are shaped in large measure by the differential spread of the resources they bring to these games.

Other classical discourses agree that states are powerful, but seek instead to itemize their vulnerabilities (for example in relation to the unmet concerns of domestic publics, or definitions of human needs) and to design schemes to restrain their excesses. Grotian modest proposals, of states acting cooperatively within a context of cautious definitions of international community, are part of this fabric of ideas. Kant thought harnesses of international government are feasible, especially in the context of constitutional change within states, despite what he acknowledged as the pressures on states in a condition of anarchy to expand their power (Huntley, 1996: 55–6). Various means have thus been proposed to weaken or sever the grip of powerful states on the structures of world order. These include building international law, encouraging the incremental erosion of the barriers to humanitarian intervention, cultivating the work of transnational civil-society groups, empowering smaller states through regionalism and constitutional change in IGOs, and facilitating multi-actor coalition-building and consensus-formation in relation to specific global issue areas. Environmentalism has been a beneficiary of each of these strategies.

States are variously bolstered or routed in different discourses. In the functionalist argument of Mitrany and others, states acting alone or even in concert with other states cannot fulfil the needs of citizens. An expanding web of task-oriented cooperative activity centring on multiple kinds of actors becomes instead the basis of transnational problem-solving. Many other commentaries have remained equivocal about states and state-centred organizations. These have increasingly been judged according to the criteria of democratic governance upheld within states. Woods has identified the key emergent principles of good governance at the global level as participation, accountability and fairness (1999: 43–6). A historical irony is that the expanded roles and influence of IGOs in the

1990s in security and economic areas often underscored the continuing power of the richer industrialized states.

Claims and models

Several types of actors in global society claim important governance roles. I will look at the environmental significance of the claims of civil-society groups and states, and those associated with hybrid mixes of actors.

Civil-society models

Much environmental discourse draws on ideas of society or community at levels transcending state boundaries. While there remain large areas of disagreement about the relative empirical or prescriptive significance of different types of actors in global society, a traditionally common focus of environmentalism has been on civil-society organizations. The numbers of NGOs generally grew significantly in the last century. Boli and Thomas (1997: 171–3) identify a world polity centring on NGOs, and that is constituted by these through cultural practices. There are emergent properties in the evolving system, that is, so that it cannot be considered as reducible to discrete actors. The conception has similarities with the broader notion of a global civil society, which has been used extensively as a theoretical foundation for empirical research on environmental, human-rights, and a wide variety of other NGOs. It 'sets a more demanding standard for the evaluation of transnational political processes than has been applied in prior accounts of such activity' (Clark et al., 1998: 1–2). A conception of global civil society is also central to attempts to characterize changing governance practices at the global level (Woods, 1999: 41). Thus structural and procedural features of world society, including the ways in which authority and power are attributed to states, are 'constructed and propagated through global cultural and associational processes' (Meyer et al., 1997: 144–5). The bases of global society stem not from states, that is, but from the social, cultural and economic relations of diverse actors that themselves constitute the world of states.

Environmental NGOs are diverse. They variously focus on combinations of local, national and global concerns. Some have across-the-board environmental-policy agendas; others are highly specialized. They vary in size. The dozen or so largest US organizations have a combined membership of around 11 million (Raustiala, 1997: 731). There is also considerable variation in the degree to which groups have an interest in governance questions. The influence of those that do depends on factors such as funding, membership bases, the organizational skills of executives, lobbying skills and tactical repertoires, and the constitutional and legal settings of states, including the taxation rules of different jurisdictions. Organizations are also affected by processes of generational change and cohort replacement (Whittier, 1997).

Identifying these as actors involves more than a taxonomic exercise. Prescriptively oriented accounts of global environmental problems tend to view them favourably as governance actors. The spread of sustainable-development discourses thus acts among other things to legimitize NGO involvement in governance arrangements. They are seen variously as counters to state or MNC power, or as rational problem-solvers in areas where states are criticized as ineffective, or as compromised by self-interest or the felt need to appease powerful economic organizations. Often neglected in sympathetic appraisals, though, is the extent to which NGOs can be viewed simply as groups with traditional kinds of interests. These interests include survival, the retention of membership and funding sources, and the protection of niches and reputations. Environmental NGOs, that is, like any other organizations, make choices that reflect their interests in the light of the resources they have at their disposal. And these choices do not necessarily have good environmental consequences, any more than do those of states (Raustiala, 1997: 720).

A central issue in NGO governance claims, especially for their critics, is their democratic character and representativeness. NGOs generally share liberal-democratic values, particularly since the principle of rational voluntarism on which many depend requires this kind of value-system (Boli and Thomas, 1997: 180–1). Further, and regardless of the nature of their internal electoral or decision-making procedures, NGO policies and programmes can be seen to reflect the preferences of members and clients since withdrawal of support, for example, has practical financial and political implications that reflect judgements of organizational performance. Effectiveness and technical rationality have also been implicitly redefined as key criteria of political legitimacy. Many NGOs can draw on extended networks of scientific expertise, mostly also voluntary, to shape and defend programmes and persuasion attempts, and hence can argue they have unrivalled access to unbiased data and to findings uncontaminated by state interests.

Their case nonetheless tends to be weaker in relation to the dual question of, first, whether NGOs collectively can adequately reflect the views of larger environmentalist constituencies and, second, whether these constituencies in turn are more-or-less accurate reflections of national-society mosaics. On the former issue, the ability of NGOs to muster domestic support for environmental policy options reinforces their claims. They can bolster what might otherwise be fragile threads of merely formal environmental agreements among states. And in practice many NGOs, and supporters who are not members, share the view that their roles in governance arrangements are irreplaceable (Clark et al., 1998: 21). In relation to the latter issue, however, NGO claims tend to collapse the issue of representativeness into arguments about the rightness of specific (environmentalist) viewpoints.

As central players in conceptions of transnational civil society, environmental NGOs have dual guises. First, they act like pressure groups in domestic politics. They try to influence government structures (national and interstate) when these deliberate on policy and draft environmental laws and regulations. Much of this activity is centred around interconnected tasks such as agenda-setting, media-manipulation and the mobilization of publics. Although important, this image misses or even wilfully excludes significant features of NGOs considered as social organizations. In the second capacity, they are important because of their autonomous activities in relation to the environment. Environmental organizations typically mount programmes of varying degrees of ambition, complexity and efficacy in areas such as the protection of habitat, environmental education, data-collection and community development. As governmental structures for environmental policy have become increasingly truncated in the North as a result of deficit-management and other pressures in the 1990s and early 2000s, activities such as these have become increasingly significant. The number and variety of partnership or joint-venture schemes between governments and environmental NGOs have grown accordingly.

Partly as a result of these interactions, debates about environmental-policy distinctiveness or exceptionalism have been a sporadic feature of discourses. The argument is also a tactic for underscoring the perceived urgency of environmental issues and legitimizing NGO participation in governance arrangements. The particular features include the scientific and technical character of many of the sub-issues, and the global-commons aspects of environmental questions generally. These and other characteristics have also supported advocacy of national and global governance practices differing significantly from those in other areas. However, comparative exploration of issue-areas reveals limitations to this approach. NGO activity is common to other areas such as human rights, just as technical complexity characterizes nuclear-proliferation and other issues. Methodologically a more useful cross-area approach – provided discrete policy areas can indeed be defined – is to identify the distinctive features of each. This kind of exercise is central to functionalist ideas of governance, which will be discussed later.

There are different framings of the central questions of enquiry in relation to these topics. Environmental groups can be studied primarily because they tell us something about NGOs in general terms, or because they might reveal important information about the handling of environmental questions. Typically, both research objectives are present, but one or other tends to predominate. From the first perspective the limitation of studying environmental questions is that these comprise merely studies of a single case or issue-area (Clark et al., 1998: 1). An understanding of environmental NGOs thus requires comparative assessments of groups in other transnational sectors such as human rights, education, or agriculture. The second approach can contribute to these comparative goals, but it is valuable primarily because of, not despite, the relative discreteness of environmental-policy issue-areas. An initial civil-society-based image of environmental governance thus captures NGOs within a flexible, pluralist taxonomy of state and non-state actors. This approach relaxes for the moment assumptions about the priority to be attached for purposes of analysis, normative discourse or policy development to certain types or groupings of actors, whether states, science-based epistemic communities or others.

The 'foreign policies' of NGOs in these settings are directed mainly towards two types of relationships: with other NGOs, and with states. Firstly, NGOs are typically situated in complex networks of relations with others. Many share similar objectives. But they also compete for resources, including supplies of rational voluntarism, and for niches in environmental policy space. As Gale writes, 'global civil society is not a realm of free, harmonious asssociation' (1998: 347). Competition may be accentuated by stereotypes of particular organizations as too staid, too noisy or too technical. In relation to the Antarctic environment, acid rain, tropical forests and other questions, collaborative inter-NGO arrangements have nonetheless been common. There are cross-area extensions with groups in overlapping regions such as human rights and threatened indigenous peoples. Coalition-making activities on specific governance issues are shaped, like those of firms contemplating strategic alliances with other companies (Gulati and Gargiulo, 1999: 1440), by uncertainties concerning the capabilities, resources and aims of potential partners.

Secondly, civil-society groups have to make decisions about their preferred relations with governments. The spectrum of choice includes participating in the making of unsatisfactory compromises, and engaging in extra-institutional protest that risks their marginalization (Riddell-

Dixon, 1999: 168). Dilemmas are made more acute by the self-identification of many environmental NGOs as members of a social movement, particularly of one of the 'new' versions that deal with issues of gender and human rights, and problems of community and identity (Barker and Dale, 1998: 70–2). A social movement has been defined in general terms as 'a collective effort to change the social structure that uses extra-institutional methods at least some of the time' (Minkoff, 1997: 780). Both reformist and radical environmentalists may be compelled to trade off social-change with environmental goals, though the two are normally inextricably mixed in complex means–ends equations.

State-oriented and inter-NGO activities are closely linked in practice. In multilateral forums, for example, NGOs lobby state delegates and members of IGO secretariats, and also create parallel inter-NGO frameworks for promoting discourses and actions (Clark et al., 1998: 19, 24). These issues have increasingly been shaped by problems of defining relations with MNCs. During the late 1990s corporations planning investment in the South began increasingly to consult with environmental and other NGOs.

State-based models

States have traditionally been central to environmental regimes. International environmental law has expanded considerably since the early 1970s (Burhenne, 1997; Kiss and Shelton, 1997: ch. 5). There was a corresponding growth, at least during the earlier part of this period, in capacity-building by national governments, for example in the creation of environmental-impact laws and environmental agencies (Frank et al., 2000: 98–9). States have become more significant players since the transition in the late 1980s and 1990s to different types of governance arrangements in areas such as ozone-layer protection and climate change. The two interrelated sides of this transition are that such undertakings eat much more deeply into national economies and sovereignty than did the sectoral and technical environmental agreements of the 1970s; and that states are closely involved in their negotiation as matters of higher national priority (Raustiala, 1997: 732; Peterson, 1998: 416–7).

States also operate discrete environmental regimes within territorial jurisdictions. Without their capabilities and political will, international environmental agreements can neither be created nor implemented. These agreements are shaped by the implications of sovereignty. The Convention on International Trade in Endangered Species (CITES), for example, focuses on endangered-species issues connected with international trade, and for the most part steers around non-trade-related conservation questions that fall under the domestic jurisdiction of states. States extend environmentally significant policies across boundaries in the form of development-assistance programmes, trade arrangements, and signals of support for or resistance to emerging environmental rules. States possess coercive and regulatory mechanisms which NGOs occasionally want to exploit, though, as in the case of anti-nuclear-waste protestors in Russia in the 1990s, they also turn these powers against environmentalist causes. Their active support is crucial for many NGO-based initiatives, for example the 1997 land-mines agreement. States are increasingly cast in the role of defenders of environmental-policy and social-welfare fabrics threatened by globalization. Their supporters have argued that states are likely to be more alert to threats to natural resources, as Iceland was over its fisheries in the 1950s, than is possible in more extended forms of transnational governance. States are often the main, though not the only, actors in responses to environmental disasters, such as the early 2000 cyanide spill on the Danube. State-based scientific organizations in OECD countries have been crucial in initial stages of problem-definition, for example in relation to ozone-layer decay over the Antarctic in the 1980s, and the discovery of possible transportation mechanisms of East-Asia-sourced pollutants across the Pacific in the late 1990s.

Claims vary among states, however, and some have been eroded by change. Defence of the principle of ecological sovereignty has become rarer in the North since the 1980s. The smaller West European states have increasingly maintained that this does not serve the interests of citizens. The argument does not so much undermine the state, however, as shift the focus to the level of regional interstate arrangements. Further, both sub-national and external extensions of the state have become significant in evolving environmental governance arrangements.

Firstly, states have a variegated internal character. Depending on the constitutional and other features of particular states, government environment agencies have been able to take on important external policy roles, sometimes in an uneasy competitive relationship with foreign ministries. This is part of the larger process by which domestic policy areas in Western states have increasingly become internationalized and subject to constraints within the G7, WTO and other groupings. The EU represents a special case, in that it has historically rested on the creation of regional governmental institutions of an evolving quasi-federal character. The workings of the principle of subsidiarity, as well as the parallel rise of regions, have also expanded the sub-national element in EU policy-making. Linkages between domestic and international governance processes have particular importance in federal systems. In Canada and Australia, significant authority in relation to environment and resource issues rests with sub-national governments, while foreign-policy authority, including the negotiation of international agreements, remains with central governments.

A second set of factors crucial to this modifying process has been the changing role of intergovernmental institutions. IGOs vary from all-purpose and universal entities such as the UN, to specific, sectoral, and technically-based organizations such as the World Health Organization. Many have developed environmental programmes derived from interpretations of their larger mandates. IGOs considered as actors, however, are constrained by the states who provide their memberships and sources of funding and often their bureaucratic personnel, and who have the last say on programme and policy matters in their executives and in the resolutions passed by their assemblies. IGOs are thus in many ways forums for interest-based interstate cooperation. As actors, they have to adjust their activities according to their interpretations of what states want. But they also have considerable room for political manoeuvre. This capability arises from prevailing uncertainties about goals and strategies, the momentum or inertia of past policies, member-states' own wish-lists and their requirement of being able to provide domestic audiences with documented evidence of IGO action and efficacy, and, to some extent, the attractiveness for states, in terms of prestige, of being able to demonstrate that they are contributing to IGO goals. Such processes have gained in importance as more emphasis has been placed on consensus-formation in IGO decision-making. IGOs can thus attempt to control settings by, for example, '"forecasting" the need for their services and the nature of their clients' (Haas, 1990: 56).

However, states have also changed in more fundamental ways. In their traditional Westphalian guise, they were conveyers of the identities of persons, solvers of problems and satisfiers of needs. Processes of economic globalization have eroded many powers, and changed the expectations of domestic actors. As a result of migration, transnational social spaces facilitate the building of civil-society groups across national boundaries (Faist, 1998: 223, 244). Secular value changes and other factors, some enhanced by responses to globalization, have produced a diversity of creolized, negotiated and other identities that contrast with traditional features of nation-state-based identity (Hedetoft, 1999: 76).

The governance claims of IGOs considered as autonomous actors are constructed on the basis of these and other alleged weaknesses of states. Lack of direct enforcement power is not necessarily decisive in all cases.

IGOs have other instruments, including moral embarrassment, the possibility of persuading states to exercise enforcement functions, and authority arising from their roles in state reporting procedures on international agreements, such as the meetings of parties of CITES. IGO-sponsored events often have significant consequences for states. The Stockholm environment conference of 1972 led many states to expedite the creation of environmental agencies and national legislative frameworks (Boli and Thomas, 1997: 186). The Convention on Biological Diversity (CBD) of 1992 stimulated the growth of habitat-protection and species-recovery programmes in many Western states. Indeed, the analysis by Frank, Hironaka and Schofer suggests that IGOs are the prime movers in top-down processes of environmental-policy change in global society. In environmental policy, 'to a surprising extent, the blueprints for nation-state involvement are drawn in world society, from where they diffuse to individual countries'. The process occurs in other policy areas, they note, 'but it is especially prominent in environmental protection, where laws and problems seem to flout national boundaries'. Thus 'nation-states environmentalize in response to global institutionalization' (2000: 96, 99, 111). However, the presence of an enforcer has been argued in game-theoretic analyses to be a crucial requirement to prevent actors free-riding or reneging on international environmental agreements (Sandler, 1997: 29–32). IGOs remain severely constrained in their capacities to perform these functions.

Functionalist defences of IGO claims, especially those of technical or scientific organizations, maintain by contrast that states are either hazardous to their citizens' interests, or at best awkwardly necessary facts of life. Related defences, particularly in relation to the environment and oceans, hinge on the concept of a global commons or stress the common transnational elements in specific problem-areas. IGOs also have critical roles in relation to the generation of non-partisan data, and in procedures for monitoring state observance of environmental agreements.

In addition to frameworks that point to their role in producing public goods in a world of states, IGOs are also central features of embedded-liberalism models. IGOs here respect some features of sovereignty, including the qualified autonomy of the state in relation to domestic policy areas, while defining or helping to enforce rules that prevent disputes from erupting into intense and protracted economic conflicts (Dunoff, 1999: 734–5). Under pressure from a mix of IGO and other external expectations and domestic mobilization, OECD states gradually greened themselves from the early 1970s. At the same time, a crucial dynamic centred on mechanisms to ensure that the uneven spread of state-based environmentalism did not generate differential economic advantages or vulnerabilities. Incentives, that is, should not be given to jurisdictions to defect, effect weak environmental rules as elements of strategies to attract foreign capital, or use environmental rules as non-tariff barriers. Environmentalism has thus become a critical element of the harmonizing fabric of Northern trade regimes. However, the linkage also exposes IGOs to criticism for failing to recognize the environmental and social costs associated with projects and programmes encouraged in the South.

In both public-goods and embedded-liberalism accounts, the nature of the impacts within states of IGO activities remains a crucial issue. International environmental arrangements vary widely in the range and depth of their impacts on national regimes (Weiss and Jacobson, 1998). The EU contains a regional environmental regime with far-reaching policy implications for member states, especially since the inclusion of environmental policy within the common pillar defined in the Maastricht and Amsterdam agreements of the 1990s. Despite the readiness of the European Court of Justice to act against non-complying countries, however, considerable divergence has persisted in the ways national governments respond to EU regime developments such as those on wild birds and habitat protection. At the other end of the environmental governance spectrum are much looser agreements in which even tacit powers of compulsion are severely constrained. Compliance problems are frequently anticipated in multilateral negotiations, however. This factor shapes the positions and negotiating strategies of states on convention provisions such as exemptions rules and time-scales for implementation. Post-agreement domestic politics have nonetheless become increasingly turbulent, for example as a result of the objections of oil companies in the late 1990s to the terms of the 1997 Kyoto Protocol on climate change.

Hybrid models

The closeness of interstate environmental arrangements to politics at domestic, even local, levels, however, means that confining attention either to states and their creations in international law and organizations, or to civil-society groups, seriously limits both analysis and prescription. Many arrangements have a mixed character. Critical issues surrounding the participation of some actors, particularly corporations or institutions representing their interests, remain unresolved.

Despite limitations on their access to the interstate world, environmental NGOs have become central to transnational governance processes through data collection, the monitoring of state behaviour, public mobilization, and in other ways. As noted earlier, they have developed a variety of cooperative relationships with both states and IGOs. North–South projects characteristically comprise multiple actors, for example Costa Rica's participation in nature–debt swaps from the 1980s and in carbon-bond trading based on its forest carbon sinks from the late 1990s. Some NGOs reject this role as a politically unacceptable co-optation exercise, or define their own activities as being 'non-political'. Some, like the Worldwide Fund for Nature (WWF), actively promote diverse partnership and collaborative arrangements with national and interstate bodies. Greenpeace, as a highly decentralized transnational network of national and local organizations, is associated with mixes of strategies that combine active and highly publicized opposition to companies and state-based entities, with cooperation with states and IGO secretariats in multilateral forums. The World Conservation Union/IUCN has traditionally included NGOs, state agencies, and states as members. It has operated both as an NGO in pressuring IGOs into action, for example in relation to CITES and the World Charter for Nature, and as a quasi-IGO in monitoring and facilitating the workings of international agreements. The InterAction Council, an NGO, consists of former heads of government and state. It has included sustainable development and protection of the natural environment among the principles it wants to see included in a UN-sponsored Universal Declaration of Human Responsibilities (Westing, 1999: 157). Many transnational scientific organizations have sought out niches in arrangements such as the Antarctic environmental system, while others, and individual scientists, have preferred the more openly critical politics of traditionally structured environmental NGOs.

There is thus considerable diversity in the patterning of governance relations among NGOs and state-based regimes. NGOs, as noted earlier, collaborate with states in multilateral settings (Clark et al., 1998: 13–17). Their ability to do this rests partly on internal organizational attributes. They also require funds for travel or for maintaining offices in IGO headquarters. They depend on diverse resources such as the political skills of individuals, and discourses that construct environmental questions in transnational and civil-society terms. But their actions are also contingent on a facilitating political environment. States may agree to include NGO members on delegations to IGO assemblies, or as participants in exchanges of views in informal extra-conference settings. Environmental, human-rights and other NGOs expanded their participation in UN conferences in the 1980s and 1990s (Fomerand, 1996: 362–4). UNESCO is among several organizations that have traditionally defined a variety of consultative relationships with NGOs in sectoral areas which give the latter access to Assembly and other organizational processes.

States are increasingly in a relation of dependence on NGOs for data, or assurance of domestic support for negotiating positions at multilateral conferences. Government environmental agencies, like others weakened by deficit-cutting measures, in practice have to nurture supportive external constituencies to acquire tacit allies in internal battles. Partnership arrangements between environmental NGOs and governments have to some extent accommodated the partial retreat of many OECD governments from traditional post-1970s environmental-policy areas. The WWF in Canada, for example, has provided funding for and participates in the mixed government-NGO scientific work that determines the threat status of wildlife species. NGOs are not merely sources of outside pressure on governments, then, so much as increasingly attractive collaborators in schemes that government agencies and IGOs wish to support (Raustiala, 1997: 720, 724). One important role has been in the monitoring of international agreements (Fomerand, 1996: 368). The data-gathering work of TRAFFIC, a UK-based NGO, is a key feature of the continuing politics of CITES implementation. For NGOs, many such activities mix political and policy goals. They can support and take part in interstate deliberations while simultaneously seeking resources for anti-state lobbying and campaigns. Governments, though, have multiple relations with domestic actors in addition to environmental NGOs, and in large measure are able to define the circumstances in which cooperation with these is welcomed or spurned.

Regional environmental governance is also shaped by traditional practices, for example the customary sea tenure arrangements of the Solomon Islands. These, however, have been increasingly restructured by economic developments at regional and global levels, and are affected by civil strife. As a result they may eventually dissolve into open-access regimes (Aswani, 1999: 448–9). An increasingly important group of transboundary environmental and resource-management questions, with implications for both states and international institutional arrangements, is posed by the governance claims and autonomous activities of aboriginal organizations in Northern states. Management practices based on customary rights and land claims are a growing feature of domestic regimes in the USA, Canada and Australia. In New Zealand, definitions of Maori principles of resource management have been incorporated into the constitution, but government conservation objectives have clashed with indigenous use practices in relation to some wildlife species (Chanwai and Richardson, 1998).

Resource, chemical, pharmaceutical and other companies have been associated with emerging governance practices in some areas. There are institutionalized forms of cooperation with MNCs and industry associations, for example, in some sectors of the work of WHO, and in the food-additives and environmental standards work of the Codex Alimentarius Commission. However, extension or greater formalization of such practices remains problematic. There is a lack of consensus both on industry participation in principle and on the forms it should take. This is despite the power of MNCs in the global economy and in relation to environmental issue-areas. Foreign affiliates of transnational corporations accounted for worldwide sales of $US 9.5 trillion in 1997, compared with world export figures for goods and services that year of $US 6.4 trillion (Hedley, 1999: 216–17). The issue for governance, though, is not the fact of MNCs' power, but their positioning in relation to the multiple problem dynamics of environmentalism. The environmentalist, and still largely the state, preference tends to be to rely on mixtures of IGOs, civil-society actors and states in governance arrangements. States in such settings are thus insulating or mediating mechanisms between systemic economic forces on the one hand, and individuals and communities on the other (Breen and Rottman, 1998: 10–11).

Functionalist arrangements

The role of the state is a question that has traditionally confronted governance designers at any level, whether the focus is on transnational arrangements or local-community stakeholder or public-goods formats. The anti-state position was traditionally favoured in classical IR accounts by advocates of authoritative IGOs. This view was rejected, though equivocally, by other liberal critics of the state system. The functionalist argument called instead for the participation in transnational governance of different sets of actors, including states, depending on the specific problem being handled. Variations of this latter argument continue to have practical implications for the design of environmental governance arrangements. Like traditional approaches to international environmental law, the functionalist critique draws on the observation that many pollution, species-conservation and other environmental problems have significant transboundary dimensions. Unlike these approaches, it rejects the inference that cooperation among states is a sufficient, or even a necessary, basis for solving these problems.

The functionalist argument in IR was associated with writers of the interwar period, primarily Mitrany, who aimed to produce concrete plans for a more peaceful international system (Boardman, 1999). A practical concentration on discrete tasks was the preferred means, with collaboration among a variety of state and non-state actors as the primary mechanism. Related ideas have resurfaced in debates on environmental and other governance problems in the post-cold-war world. They connect with discourses on social-welfare problems under conditions of deepening globalization. Functionalist precepts also have a practical appeal that rejects the constitutional-blueprint approach of traditional world-government theorists. Functionalist arguments stem, too, from recognition of the facts of economic interdependence that increasingly preoccupied writers of the 1920s and 1930s (Sterling-Folker, 2000: 101).

The agenda of functionalism, however, remains that of peace. This elusive condition tends to be viewed in functionalist accounts as the by-product of transnational collaboration among technically competent actors on practical matters. While the two goals – peace on the one hand, and governance in relation to concrete tasks on the other – are not inconsistent, the characteristic functionalist juxtaposition of these raises questions about the value of the approach considered simply as a way of handling problems. Design problems of global governance, that is, are viewed not only from the perspective of technical rationality or public-administration or civil-society efficacy, but also in terms of the criterion of their indirect contribution to peace-building. These sets of tasks can more usefully be considered as analytically and prescriptively separable. Some environmental issues, however, are components of both sets.

Environmental issues have a particular resonance with this approach to the management of issues in global society. Firstly, some were increasingly defined in security-related terms in the 1980s and 1990s. These included problems in debates on sea-level rise resulting from anthropogenic climate change and its effects on coastal and small island states. Environmental issues thus became tied into the characteristic functionalist linkage between peace and security on the one hand, and technical problem-solving on practical matters of human needs on the other. Secondly, many environmental issues have transnational aspects, like the spread of disease vectors and locusts in East Africa, acid precipitation in North America and western Europe, or the transport of a range of pollutants by air and water among contiguous regions. Many of these have local transborder dimensions, for example the US-Mexico pollution problems surrounding the San Diego area. The fluid mix of multiple stakeholders that characterizes many environmental problems, then, makes them especially approachable by means of classical functionalist logic.

This logic took various forms in the period from the 1920s to the 1940s. Functionalist variability reflects the diverse philosophical influences that are detectable influences on its development. Significant elements include strands of nineteenth-century (reformist) socialism and utilitarianism. A certain utopianism is present too in the functionalist hunch that politics, in its bad forms, can somehow be avoided. People and organizations are seen to have a capacity to collaborate and solve common problems without it arising. There is often 'a clear need to work on practical problems together', but some 'political or other constricted interest' gets in the way (Mitrany, 1975: 45). There is also a streak of incrementalist caution and even conservatism at times in Mitrany. He develops Burke's argument about the need for governments to concern themselves with practical matters (1975: 113). He also refuses to join the anti-state critics who have at various times since the interwar period wanted more frontal assaults on state power, whether through the design of authoritative international institutions or through radical domestic social change.

Instead, the approach builds on linked assumptions of both the discreteness and the interconnectedness of social problems. First, problems are separable. Each can be tackled as such in governance arrangements. The functionalist approach helps to 'bring out the real elements of whatever issue is at stake' (Mitrany, 1975: 45). Thus activities or tasks are to be 'selected specifically and organized separately, each according to its nature, to the conditions under which it has to operate, and to the needs of the moment' (p. 116). Secondly, problems are viewed as interconnected, just as they are seen to be in classical environmentalist accounts. Cooperation in one area creates a need for cooperation in others. Difficulties and gaps encountered in one lead to fresh approaches that highlight the design problems of others. This dynamic process produces patterns and experiences of cooperation among actors that result in the emergent properties of a governance system designed primarily from the bottom up rather than in top-down fashion. International government thus becomes more and more 'co-extensive with international activities' (1975: 112–13). Form, in the classic functionalist expression, thus follows function.

These concerns, moreover, operate across the different levels of analysis that were increasingly being segregated in realist IR frameworks from the late 1930s. Thus Mitrany maintains that there is an 'identity' of governance objectives, whether we are dealing with the domestic or the international level. These are 'to create equality before the law for all members of the community, that is, to establish legal justice', and secondly 'to create fair material conditions of life for all those members, that is, to promote social justice' (Mitrany, 1933: 103). This wide-ranging cross-level interest is closely tied to Mitrany's identification of pragmatism as the primary basis for selecting the actors appropriate for problem-solving in any particular area.

Here we encounter an ambivalence in Mitrany that reflects conflicts in the peace debates of his day. Thus he argues in effect at different times both that the state is a dysfunctional actor and also that, depending on the issue and the circumstances, it has good claims to be considered a practical problem-solver along with non-state actors or IGOs. Much of his writing on functionalism does indeed have an anti-state tone that responds to more far-reaching critiques of state capacities in relation to human needs. The preference is often for problem-solving by technically-grounded intergovernmental agencies such as the World Health Organization. However, he is carefully permissive in other, particularly in earlier, statements about the range of actors to be accommodated in governance arrangements. Thus, 'No one can be certain that our needs demand that we should discard the State altogether . . . To assert that outright would be to fall into the same error of a priori dogmatism which we deplore in the opposite schools. All that we need do is to approach the problem with an open mind, keeping our gaze steadfastly on the end rather than on the means' (Mitrany, 1933: 98).

States, then, can in functionalist logic be a means towards the end of environmental problem-solving. They are not necessarily the optimal actors; but nor, from this perspective, can any actor be viewed on principle as the most suitable. Functionalist logic would also veto many of the governance claims of environmental NGOs. Or, more specifically, these claims would be subject to re-examination in relation to each issue. The participation of actors is thus open-ended and subject to continuous re-evaluations of needs and activities. At the same time, though, it is also evident that Mitrany, like other functionalist writers, had institutional preferences for inter-governmental technical bodies in areas such as health, agriculture and other fields, of the kind that had emerged first in nineteenth-century Europe and which developed under the League of Nations and then the UN.

Functionalist advocacy of a pluralist approach to governance, flexibly designed in relation to specific problem-areas, thus tends to leave open many questions of actor participation and decision-making. Partly this strategy is deliberate. However, some critical gaps are left. Firstly, the principle of governance based on the participation of actors capable of handling problems in particular areas requires issue-definition processes. The environment as such a potential issue-area has arguably become too broad and conflicted to constitute a foundation for task-based governance. It has serious limitations as an organizing device able to draw discrete issues together for these purposes. Functionalist governance breaks tasks down further, for example to the problem of the quality of potable water or regional migratory-bird frameworks. An organizing and explicitly 'environmentalist' framework might accordingly disappear, with profound consequences for the politics of many of its sub-areas. Specific sectoral schemes organized along functionalist lines may thus lead to fragmentation and a lack of coherence in environmental policy-areas. UN framework agreements, such as the UN Law of the Sea convention, can work around this problem by calling for the setting up of associated subregimes. This still leaves open, however, the question of the possibility and desirability of directive pushes from global IGOs in relation to large clusters of issues in a decentralized governance system.

Secondly, actors in a sense define the bases of their own roles in relation to functionalist arrangements. NGOs (like states earlier) have traditionally had a hand in shaping the discourses that consolidate their own participation in governance. Debates over participation rules echo older arguments made by functionalist critics. These can be seen, for example, in the contrasts between the approaches of traditional state-based regimes to environmental and resource issues, and those of property regimes in which the multiple users of a resource become the key players in sustainability debates and practices (Princen, 1998: 395–6). There are comparable issues in stakeholder approaches to environment and resource dispute-resolution and policy-development processes in national and local contexts. Both levels reflect the increasing attention paid during the 1980s and 1990s to concepts such as democratization and participation in emerging definitions of sustainable development. Assumptions of the growing inseparability of these levels can also be seen as reworkings of basic functionalist premises. A related avenue of facilitating discourse for NGOs has been through broader epistemic arguments about the character of environmental and resource issues. These stress the knowledge-based elements of questions, and with the aid of problem-solving metaphors serve to legitimize the roles of data-gathering NGOs.

Companies are more perplexing for functionalist logic. These are powerful shapers of the economic and social contexts within which environmental decisions are made by governments and civil society groups. National governments, like other actors in transnational governance arrangements, cannot be seen in isolation from these larger economic settings. MNCs raise issues in their relations with host governments that have a bearing on global governance. These include questions about the capacity of states to enact regulations on environmental standards, as well as problems that indirectly have a bearing on environmental policy such as employment, technology use and development, competition and trade, and fiscal policy (Hedley, 1999: 218–19). They promote and benefit from specific government policies, though there is considerable variation as between the US and UK, continental European and Japanese models in definitions of the role of their home state (Breen and Rottman, 1998: 6–7). Firms have also become increasingly prominent in environmental governance activities at national levels, for example in the design of voluntary codes, and in approaches to trade liberalization issues (Clapp, 1998: 298). States have edged away from regulatory solutions to environmental problems not only because of diminishing capacities to control transborder flows of goods and services, but also out of a growing convergence among state and private-sector interests and role expectations.

Some functionalist writers have traditionally fused anti-company with anti-state postures, for example in the campaigns of the 1920s and 1930s against the international arms trade. Mitrany, though, fails to address satisfactorily the issue of MNC participation in governance arrangements. The general principles he sets out have divergent implications. On the one hand, companies and industry associations clearly fit in as stakeholders that should have a say in evolving governance arrangements grounded on pragmatic criteria; on the other hand, if we follow Mitrany's treatment of problem-solving and interest as mutually incompatible, the fact that they are closely identified with the pursuit of economic interest would probably deny them a seat.

Governance and global ecology

Problems of governance at the level of IGOs and multilateral environmental agreements have become increasingly central to the handling of global environmental problems. Indeed for much environmental discourse, as for IR, this seems like a statement of the obvious. Governance, though, is only one of several key theoretical approaches to global environmental enquiry. Logics aimed at the strengthening of international environmental law and intergovernmental agencies often rest on assumptions about dynamic processes of social and political change, and more particularly about the ways agreements negotiated among national governments lead to enforcement mechanisms, and possibly changes of habits and attitudes too, within states. While such frameworks are state-oriented, they are not necessarily state-centric. Other actors, including

IGO secretariats, environmental NGOs and transnational scientific associations, play key roles in problem-definition, the monitoring of agreements, and the creation of support for governance arrangements among domestic constituencies. However, though there are many counterbalancing instances of spreading networks of implementation and good governance, patterns of compliance remain uneven within and between North and South. The substantive provisions of agreements are characteristically diluted as states' anticipations of compliance problems and post-agreement politics shape negotiations. Sovereignty principles affect the range and the forms of possible agreements. A relative neglect in IR of the domestic consequences of international environmental conventions reflects traditional divisions within the disciplines of law and political science, but it also highlights conceptual difficulties inherent in the apparently simple notion of compliance.

These kinds of approaches to environmental governance, constructed on traditional practices of statecraft, have been increasingly joined by other models. Some conceptions of problem-solving grounded in conceptions of global civil society share a determinedly anti-state, as well as an anti-MNC, character. Others pursue a more secure environmentalism based in part in state structures that have capacities to defend environmental public goods against both economic globalization and IFIs. The potential for effective alternative practices of governance is suggested in traditional functionalist accounts. These share with pragmatic forms of environmentalism a flexible problem-oriented approach. However, functionalist discourses and practices developed historically in relation to the problems of Northern states, and still tend to have more applicability in those contexts than in Rawls's 'burdened societies'. The politics of recognition also constrains the development of functionalist governance. The systematic extension of present arrangements beyond states to include NGOs and MNCs confronts opposition, respectively, from critics who complain of the lack of representativeness and political legitimacy of the former, and from others who define the latter as the major source of the problems governance arrangements are set up to address. Some practices, such as MNC consultations with environmental organizations and extensions of IGO consultative status for NGOs, nonetheless point towards a growing flexibility and hybridity in arrangements.

Part III
Wefts

8
From City-State to City-Planet

The discussions in the previous group of chapters looked at various theories directly, or potentially, related to global ecological themes. For the most part these made their entrance in a general-theory guise. Each broad approach was viewed, that is, in terms of its capacity to generate frameworks for dealing with the universe of global environmental problems. Though diverse, and often characterized by internal conflicts, each also has a capacity to produce evaluative judgements of the likely contributions, and typical failings, of other approaches. In some cases, as with approaches derived from neoclassical economics, this has sometimes taken the form of ignoring competing theoretical strategies while advancing into terrains traditionally associated with other disciplines.

In this chapter these perspectives will be viewed less as differing and separate routes towards comprehensive accounts of global ecological problems, and more in terms of their auxiliary roles as purveyors of basic ingredients that can be used in different mixes in different recipes. The respective metaframeworks of ecology and political economy suggest both divergent and complementary tendencies in the understanding of environmental questions. Each interweaves in a different fashion with external disciplinary perspectives, and with the ethical, individualist, governance and other traditions of global environmental commentary. They are expressions respectively of strong and weak versions of global environmental argument, depending on the nature and intensity of the explanatory significance attached to ecological factors. Firstly, ecologically grounded frameworks draw on a strong version of environmental logic. These suggest the possibility of general-theory ventures that accommodate natural- and social-science concepts to create standard accounts applicable to global and to other levels of environmental actions. Economic and social factors in such approaches are primarily conditioned by understandings of the dynamics of ecological processes. Secondly, the weak environmental argument points towards a variety of approaches to ecologically enhanced or restructured economic frameworks. In these the environmental domain occupies a more significant place, but it is nonetheless still interpreted as conditioned by factors such as the structures and productive relations of economies and the processes of economic globalization.

This chapter looks first at approaches to globalization processes and their significance for the environment, and then compares these with some of the underlying assumptions characteristic of interpretations based on research in the ecological and earth-system sciences. More specifically, I argue, firstly, that ecologically framed general accounts have tended to neglect transformationist arguments, which, in different forms, are often central in interpretations of globalization; and, secondly, that political-economy frameworks have traditionally minimized consideration of the impacts of dynamic ecological processes, and of the multiple uncertainties that characterize these.

Globalization: guises and gazes

Terminologies of globalization have been used in debates since the 1980s in different ways in a variety of societal and academic discourses (Ferleger and Mandle, 2000; Nederveen Pieterse, 2000b; Held et al., 1999; McGrew, 1998). Woven into these multiple perspectives are contrasting notions of cause-and-effect relations among key variables, including conflicting views of ecological processes and their significance in relation to economies. Debates also comprise normative judgements about processes and outcomes derived from the prior values of observers about markets and governments, communities and individuals, and ethnicity and cosmopolitanism. Accounts differ as to whether the phenomenon is novel or a contemporary reworking of old stories. There are many analytical starting points, from the mechanisms by which capitalism works to the ways different types of social, economic and political actors orient themselves to changing worlds (Mittelman, 1994: 427). Strategies of resistance to economic globalization by civil-society actors or revitalizing states at the same time constitute different forms of globalization redefined in social, cultural or political terms. Globalization dissolves and re-forms under the gaze of observers. There are 'perhaps as many meanings as there are analysts' (Sjolander, 1996: 603).

Enquiry into the implications of globalization for environmental change thus depend crucially on the theoretical perspective on globalization being adopted. Further, the range of processes included under the heading of globalization sometimes acts to undermine the analytical power of the concept. Like 'environment' in the hands of other critics, it buckles if it is asked to explain too much. Globalization effects should in principle be analytically separable from those of other factors. Three broad sets of views of globalization can be discerned, each with diverse implications for ecological problems. I will look briefly at economic theories, plurality perspectives, and approaches emphasizing complexity.

Economic bases

First, globalization signifies a process of transformation in the way economies operate. Each aspect – the focus on process rather than condition or end-state, and on qualitative rather than incremental change – is important. The critical feature is the growing transborder context of economies. The global economy created or significantly shaped by information technologies has a planetary scale and, as Castells (2000a: 9–11; 2000b) argues, the differential capacities of actors to generate information redefine productive capacity and relations. The most prominent systemic change has been the appearance of the firm in an increasingly multinational guise, a process of change that began in the late 1960s. The MNC in a sense has a place equivalent to the state in traditional IR frameworks, that is, as the primary actor in a complex system characterized by a differential spread of resources and power. Globalization in its earliest contemporary formulations, in the 1980s, comes from the literatures on international economics and commerce, and refers favourably to changing business environments and the ways these create new opportunities and challenges for entrepreneurs.

The effects of greater mobility by MNCs are multiple. Crotty, Epstein and Kelly have suggested several alternative models of these processes. In the first, MNCs can gain power over national and local communities, and successfully trade off increasingly impoverished states, labour unions and other actors. Secondly, and alternatively, the presence of MNC and local-company competition increases the quality of educated labour and infrastructure as jurisdictions compete to attract foreign direct investment (FDI). The third interpretation is the conventional neoliberal synthesis. Capital mobility, in conjunction with expanding trade liberalization and state deregulation, promotes wealth and also social well-being. Fourthly, emphasis on the uneven development of these processes highlights sharpening economic inequalities between North and South and within different regions of each. Finally, a dissenting view tends to argue that, because FDI remains a minor proportion of GDP, the effects of multinational firms in terms of standard economic performance measures or social-welfare or policy-change indicators are minimal (Crotty et al., 1998: 118–19).

This typology intersects at several points with prevailing environmental critiques of globalization. In these, MNC power and capital mobility are argued to have primarily harmful effects on environments. These consequences result from such things as increased incentives to intensive agriculture and ecologically destructive forestry sectors. More indirect effects include the *de facto* co-opting of IGOs, their shared participation in the grand neoliberal project, and the social and ecological costs in Southern states of policies or restructurings promoted or funded by IFIs. The critique also points to the environmental costs of factors such as the retreat from social-welfare and macroeconomic planning roles of debilitated state structures, the widening of economic inequities within and between national societies, and, on balance, the more hazardous milieus facing civil-society groups. Creative links are thus formed between environmentalism and labour, nationalist and other critics of the culturally homogenizing and social-injustice consequences of globalization.

The emphasis on secular processes of change tends to deflate the argument that globalization is novel, or even significant in terms of its cumulative effects (Hirst, 1997). Comparable aggregate processes marked the late nineteenth century. Interactions among core regions were already a feature of the mid-nineteenth-century global economy (Chase-Dunn et al., 2000: 79), and constituent factors such as limited-liability laws have a longer history. Different expressions of the continuity thesis stem from radical and liberal economic traditions.

In the first, globalization is 'part of an ongoing process with a long history' (Magdoff, 1992: 45). It is characterized over time by the geographical spread of productive relations and a greater degree of structural interdependence in the world economy. There are, however, significant differences between historical periods. As Nederveen Pieterse argues, globalization is not just 'imperialism or neocolonialism revisited ... imperialism was territorial, state driven, centrally orchestrated and marked by a clear division between colonizer and colonized; and none of these features apply to contemporary globalization [which] is multi-dimensional, non-territorial, polycentric ... [and] involves multiple intentionalities and criss-crossing projects on the part of many agents' (2000a: 132). Ecological costs are closely linked with these economic and social dynamics, and particularly with the structural economic inequalities produced by changing productive relations. Further, economies suffer long-term effects from systemically induced incentives to corporations to over-exploit natural resources.

Continuities are likewise highlighted in liberal and classical economic treatments. However, images of the dynamics of change differ. Globalization tends here to be the contemporary expression of the process of the freeing of markets from state-regulatory or command-and-control systems that began in the eighteenth-century revolt against mercantilism. It is associated with approaches to optimal, barrier-free, movements of capital, goods and labour at regional and global levels. State authorities nonetheless retain an important role in this standard liberal account, for example through the protection of intellectual property. Studies of economic integration influenced by this broad tradition emphasize the persistence of state-based institutional obstacles or impediments to globalization processes, such as barriers to FDI, capital controls and non-tariff barriers (NTBs)(Prakash and Hart, 2000: 95–6). Environmentalism appears in such frameworks primarily as sets of reforms that optimally have minimal effects in terms of economic-growth, modernization or trade-liberalization objectives. Some approaches overlap with the views of critics advocating more interventionist states, as in parts of the institutional-economics tradition, or, more specifically, with options such as natural-resource pricing, the harmonization of environmental-protection policies among states, and more entrenched practices of industry self-regulation.

Plurality perspectives

Secondly, globalization can be approached by emphasizing changes in the interactions among global actors. A more strictly pluralist view would tend to deny the primacy of any particular type of actor, such as states in the classical IR tradition. In addition to companies, states, intergovernmental organizations and NGOs, environmentally relevant global actors less easy to categorize include state agencies that formulate external extensions of domestic policies, transnational scientific associations, media organizations, and the governments of sub-national jurisdictions. Local-level governments are in many states focal points of environmental-policy activities such as air- and water-quality monitoring and control and land use. Cities, particularly the 'global' ones, have been identified as important sites of linkages between the global economy and the daily lives of individuals (Low, 1997: 403). They are core components of extended ecosystems. As 'heat islands', cities have far-reaching environmental consequences; and they determine patterns of energy flows and agricultural production in national and regional economies.

A plurality or actor-based perspective can be used alternatively to highlight observations based on assumptions of the relative centrality of specific types of actors, either on normative grounds or on the basis of beliefs about power relations. Thus 'globalist' terminologies are prominent in classically-framed IR accounts to refer to the geographical spread of the idea and practice of statehood. International systems vary in their degrees of globality in this sense. They vary, for example, in terms of the spread of multiple types of actors, the interdependence and complexity of the relations among these, the extent of their territorial reach across the earth's land-masses and oceans, and the sharing of society-like rules by members. While NGOs are in many ways the characteristic vehicles of environmentalist discourse and action, other conceptions of the organization of global society are important. For example, to the extent that economic globalization is seen to menace the environmental regulatory power of states, national capability-building and enstructuring options become more salient responses.

Complexity approaches

A third conception is to view globalization as a complex and multifaceted phenomenon with multiple causes and diverse consequences. Some approaches here have focused on the information-technology bases of contemporary globalization. Approaches of these kinds build on plurality models, and hence assume that multiple types of actors have varying degrees of autonomy. But they go considerably further in emphasizing the mixtures of factors, levels and processes – cultural as well as economic, individual and community-based as well as transnational – present in globalization. For example, globalization changes the institutional bases which shape the production and transmission of ideas (Collins, 1998). Associated interpretations of environmentalism point to the importance of communications technologies, for example in agenda-setting and in facilitating the transnational organization of inter-NGO anti-globalization protests.

Partly this is an exercise in the attribution of differing cause-and-effect relations among variables. Thus the economic and technological roots of globalization can be viewed, as they were in the first image, as the primary producers of factors such as the spread of Western liberal-democratic ideas and concerns with human-rights issues, changing consumer habits, and the thematics of news media and the arts. In other perspectives, the structures of discourses, cultural developments and other factors are treated as analytical starting points of enquiry. Jameson sees globalization as a 'communicational concept, which alternately masks and transmits cultural or economic meanings' (1999: 58). Individuals, now part of what were formerly 'distant events' (Giddens, 1991: 9, 27), both affect and are affected by changing work environments, processes of identity-formation, and productive relations.

Complexity approaches encourage notions of spatiality and diffusion as distinguishing features of globalization. Indeed Robertson argues that some of the analytical difficulties of globalization have arisen from a confusing conflation of the term with globality. His preference is to see the latter in spatial terms as 'the general condition which has *facilitated* the diffusion of "general modernity"' (1995: 27), that is, of the phenomena that include processes of economic globalization. Geographical conceptions are similarly a feature of Beck's discussion of the expanding interconnections among what were formerly separate local and national societies (2000: 80). One result, McGrew writes, is 'a significant shift in the spatial patterns of human social organization and activity towards transcontinental or inter-regional networks of relations, interaction, and the exercise of power' (1998: 301). Understanding environmental change in these varied contexts, then, requires expanded notions of globalization to include consideration not only of multiple factors, but also complex social-spatial processes. We have to move simultaneously 'up' towards a global level and its component regional parts, in the sense of the big things of trade blocs and MNC–state power relations; and 'down' towards alternate conceptualizations of globality in the sense of the geographical spread and analytical immediacy of the little things of individuals' daily lives.

Globalization and ecological order

Economic globalization undoubtedly has significant environmental consequences. Assessing these, however, requires a couple of initial clarifications. Firstly, the environmental effects of globalization, considered in the sense of growing transborder economic flows, are not recent phenomena. The internal land-use patterns of national societies have traditionally interacted with cross-border economic activities, with consequences for pollution and for wildlife species and their habitats. Stocks of beaver in eighteenth-century Canada were dependent indirectly on fur prices in Europe, and more immediately on the state of competition between the Hudson's Bay company and French traders in their relations with aboriginal suppliers (Carlos and Lewis, 1999). One factor temporarily increasing pollution of nineteenth-century English towns was the discovery and exploitation for fertilizer production of guano deposits in

South America, which reduced demand for horse manure collected from streets by households and local entrepreneurs.

Secondly, multiple factors other than the processes of globalization cause environmental pollution or other ecological stress factors. Major landscape changes were effected before modernity and capitalism (for example, the early fifteenth-century dyke-building and land-reclamation schemes in the Netherlands) or else had national-society as opposed to transnational, origins (for example, the destruction of the wetlands of the eastern and southeastern United States, and of the hills in the area of what became Seattle). If the root condition is taken to be capitalism – a causal sequence that weakens when the environmental costs of the centrally planned economies of the Former Soviet Union and Eastern Europe are considered – globalization becomes not so much an autonomous source of effects on environments, but rather an intensifier of effects, or a medium or transport mechanism for carrying packages of hazards or economic practices across borders.

Globalization processes do nonetheless exert various effects on environments. They also reinforce or check other factors that are ecologically significant. The core sets of effects are, first, those resulting from alterations in patterns of productive relations and the use of changing technologies in industrial, agricultural and other sectors; and, second, indirect effects mediated by changes such as revisions to national or global trade regimes, value-systems, cultural practices or the habits and choices of individuals in their daily lives.

The spread of industrial production in the South, for example in China and India, adds to problems of anthropogenic climate change resulting from fossil-fuel use and reliance on GHG-producing industrial processes. Intensive agricultural development in newly industrialized countries (NICs) is premised on the requirement of expanded use of agrochemicals. These create patterns of use dependence as agricultural ecosystems lose 'natural' defences, and insect or plant pests develop resistance. The spreading use of agricultural chemicals for banana production in the eastern Caribbean was accelerated in the 1960s and 1970s by British companies, which distributed free or highly subsidized pesticides and fertilizers to farmers as part of overseas development assistance (ODA) programmes (Grossman, 1998: 195). Other factors, including the increased reliance of small farmers on sources of finance and dependence on specific export markets, in turn reinforce incentives to, and the risks of, ecologically damaging agricultural practices. There are related consequences of agrochemical use in terms of the health of farmers and workers, and wildlife and domesticated species. Fish populations, traditionally used as protein sources in conjunction with rice cultivation in Southeast Asia, decline. Monocultural practices also make crops more vulnerable to pests and to extreme weather events. There are long-term implications in terms of soil loss or degradation, as in the case of intensive agricultural development in southern Brazil. These conditions, together with the removal of forest cover, constrain flood-management capabilities. Other globalization effects link agricultural development with changes in the international food trade, for example the trade in vegetables between East African farms and European markets or bananas exported from Costa Rica to North America. These patterns have structuring effects on agricultural economies and their environments in both producing and consuming countries. Some countries in Africa that became net exporters of food in the 1980s and 1990s, such as Sudan, Niger and Algeria, at the same time experience periodic crises of production and distribution of food for domestic use (Stoett, 1995: 116).

Environmental effects resulting from transborder processes such as increased FDI, however, are difficult to gauge. To the extent that environmental costs can be identified in a given context, it may not be clear whether these result from specific features of the transaction, or whether they are comparable to the effects that would obtain in any case from domestic capital investment. And foreign, as opposed to national, sources of capital in the South may in some instances have less harmful environmental consequences. Up to a point MNCs have incentives to adopt more responsible environmental behaviour, as a result of factors such as growing consumer or shareholder environmentalist pressures in their home countries, their differential possession of greater technical expertise in environmental monitoring, and their ability to provide environmental training for managers (Gentry, 1998: 108–9; Daily and Walker, 2000: 243–5).

A layer of more indirect effects of globalization on environments arises among other things from changes in institutions designed to facilitate the freer flow of trade and capital across borders. The failed Multilateral Agreement on Investment (MAI), prepared in OECD negotiations begun in 1995, was designed as a protection for firms against government regulations and practices that discriminate against foreign companies or have distorting effects on trade. Its critics have argued that there would have been significant consequences for the environmental regulatory power of national governments. In addition to these, sub-national governments in federal systems would have been limited in their abilities to engineer local environmental improvements. The MAI would have constrained, for example, state governments in the USA that restrict the sale of state lands to citizens in part on environmental-protection grounds, institute special land-use rules in relation to wetlands and other ecologically significant areas, or have rules related to recycled materials markets that could have been argued to violate investor protections (Singer and Stumberg, 1999: 9–14).

Transformations

Accounts of globalization have thus been increasingly intertwined with arguments about environmental problems. Processes of global change are transforming both economies and ecological systems. However, the metaphors of economic change and ecological deterioration prevalent in critiques of globalization tend to diverge from those of more ecologically grounded accounts. Interpretations that draw on traditional notions of adaptation in ecology and the biological sciences, for example, place relatively more stress on patterns of responses to structural features of physical environments on the part of economies and societies. The contrast of emphasis has significant implications for interpretations of the causal relations among economic, social and ecological factors. Debates within biology on the utility of differing images of global environmental change nonetheless form a potential bridge to IPE approaches to globalization, in part by way of some of the arguments about agency discussed in chapter 6.

The metaphor of adaptation is the basis of richly evocative sets of concepts in the history of many areas of the natural, and indeed the social, sciences. It is weakened in practice, however, by its association with often misleading interpretations of environmental problems generally and global change particularly. While 'adapt' is often cited in dictionary definitions as a transitive verb (as in someone adapting a tool so that it better fits their requirements), in biology and anthropology the tendency is for users to shift ground and make it an intransitive verb. Animals and plants thus adapt behaviourally and physiologically to changes in their environments, as, in some human-ecology frameworks, do societies and cultures to their resource endowments and systemic factors. A related normative dimension of the term is sometimes implied. Anthropological studies comprise attempts to relate specific features of natural environments with technologies, and with societal attributes such as belief systems and ritual practices. Some approaches have been criticized for implying that cultural features are either 'adaptive' in the sense of being functional, or else maladaptive and dysfunctional. Infanticide, for example, is at some level a rational population-control mechanism;

however, in societies that have adopted it it has persisted after changes in population–resource–environment relations made it unnecessary on these criteria (Milton, 1996: 45–6).

Adaptation, more rigorously defined, is consistent with both proactive and responsive or passive usages. The term has been used in multiple ways in accounts in sociology, criminology and other areas of the social sciences. It is central to Rosenau's foreign-policy adaptation scheme of the early 1980s, which placed classificatory emphasis on the various ways states effect both external and internal changes in response to stimuli. However, in circumstances of modernity, rapid technological change and globalization, the metaphor fails to depict the enormity and irreversibility of the environmental changes brought about by urbanization, industrialization, agrifood development and other factors. Human societies and economies, that is, are transforming environments. I prefer to use the term 'ecosation' in this context rather than 'transformation', in order to emphasize both anthropogenic transformative activities and also the frameworks of ecological rules within which these take place.

Other usages in the history of the biological sciences are more open to this proactive, transformative imagery. Some implicitly draw on notions analogous to that of agency. Von Uexküll's approach in the 1920s and 1930s focused on the meaning-creation processes inferred by observers as the basis of animal behaviour, rather than on a search for laws governing the objective interactions of organisms and their adaptations, in the conventional sense of the term, to their environments. Organisms, from this perspective, do not have an abiding and objectively observable character. They vary according to the subject that constitutes them, for example in the context of the study of predator–prey relations. In other words, 'The changeability of the object is the most important law' (quoted in von Maltzahn, 1994: 56–7).

The emphasis on organisms 'constituting' their environments in this sense is also echoed in Lovelock's Gaia imagery. This, as noted in chapter 3, draws attention to complex biosphere effects on physical and chemical earth processes. He has argued recently that the 'deepest error of modern biology is the entrenched belief that organisms interact only with other organisms and merely adapt to their material environment'. This theoretical perspective is 'as wrong as believing that the people of a village interact with their neighbours but merely adapt to the material conditions of their cottages. In real life, both organisms and people change their environment, as well as adapting to it' (quoted in Ahuja, 2000: 8).

Lewontin argues that environments have to be defined in relation to organisms and do not exist without these. Thus:

> we must replace the adaptationist view of life with a constructionist one. It is not that organisms find environments and either adapt themselves to the environments or die. They actually construct their environments out of bits and pieces ... The first rule of the real relation between organisms and environments is that environments do not exist in the absence of organisms but are constructed by them out of bits and pieces of the external world.
>
> (1995: 86–7)

In a sense Lewontin is here doing little more than elaborating on the fundamental notion of organism–environment relations that has traditionally defined the focal point of ecological research. However, what he calls the 'constructivist' metaphor emphasizes, as does von Uexküll's approach in a different fashion, the constitutive role of organisms in relation to environments. His critique is developed as part of a wider critical commentary on theoretical biology, but has particular relevance to study of the natural environments of societies, especially in circumstances of globalization. Clearly only some aspects of these environments are 'constructed' by humans in Lewontin's use of the term. We do not change major ocean currents, for example (though some hypothesized consequences of anthropogenic climate change may lead through intervening variables to polar ice-cover changes that may do this). These kinds of structures, in other words (to revert to sociological as opposed to biological metaphors of agency) are more external, objective and distant.

In transforming environments rather than 'adapting' to them in the narrower passive sense, human actions nonetheless take place in the context of ecological and earth-system processes. Critical views of this dimension of ecosation are sometimes restricted mainly to neo-Malthusian interpretations of carrying-capacity arguments, as in many of the debates of the 1930s and 1940s about population-environment relations, and the limits-to-growth studies of the Club of Rome in the 1970s. These approaches acknowledge the importance of ecological structures, but in effect tend to externalize these and so reduce opportunities for modelling interactive processes. Ecological principles also tend to be marginalized, for different reasons, in 'management' images. These treat environmental problems as inherently solvable through the application of technology, good administration, and ingenuity. There was a similar neglect of ecological factors in the arguments about 'nature's bounty' common in

US nineteenth-century accounts, and which resurfaced in the economic boom of the 1990s.

Sociologists, and other social scientists, have often been reluctant to countenance ecological-laws arguments because of fears of the possible connotations of biological determinism, a reductionist denial of human choice and agency, or pessimism about the prospects for social and economic progress. However, modes of sociological reasoning can usefully be extended, within limits, into study of the processes of ecological change. The problems are in some ways analogous to those confronting analysts of agency-structure issues in societies. Structuralist accounts, that is, have parallels with environmentalist or resource-based arguments that emphasize constraints on economic activities, or which, as in deep-ecology models, underline the dependence of societies on ecological systems. Environmental arguments echoing conceptions of agency, by contrast, highlight instead anthropogenic sources of change. The changes can be viewed as ecologically neutral or beneficial, if it is assumed that resources are either naturally abundant or that their availability is crucially determined by price mechanisms. Environmentalist critiques normally, however, view actions as more likely to be damaging. Differing constructions of agency in relation to social and economic structures are central to both these kinds of ecological judgements.

Ecological approaches tend to seek out and emphasize complex interactions among variables. Sackman's study of the history of the California citrus industry highlights interconnectedness among agricultural-ecological, economic and social factors at the level of individual workers. As he puts it, labour was 'squeezed' into the fruits, and 'the fruits worked their way into the bodies of workers'. The workers thus 'mediated between the artificial zones of the market and the organic landscapes of the grove' (2000: 29).

Studies focusing in this way on interconnections suggest the value of revisiting Giddens's reformulation of the agency-structure issue. As discussed in chapter 6, this throws light on analogous problems in the study of social-ecological processes. Ecological structures, broadly defined, like their social counterparts in Giddens's account, can be seen in the dual light of both enabling and constraining the economic and social lives of individuals. Up to a point agents effect change and 'constitute' these structures in a context of, or, in his terms, through the medium of, the laws governing their workings. Development of the ecological analogy is open to some of the criticisms traditionally levelled at structuration theory. A collapse of dualism into duality, that is, cannot be universally effected across all sets of social- or economic-environmental relations. Ecological and earth-system structures are diverse. It is often easier to see many of these, like weather patterns or natural-resource endowments, as structuring, rather than being constituted by, the activities of agents. Even so, this more dynamic approach to ecosation has the advantage of making both human activities and ecological laws more central in theoretical frameworks, and avoids the tendency to marginalize one or the other of these often found in more exclusively agency-based or structuralist accounts of the aetiology of environmental problems.

While such cross-fertilizing imagery between sociology and ecology cannot be taken too far, then, it nonetheless highlights the importance of several sets of questions in global ecological enquiry. Social-ecological approaches in this genre usefully draw on the structurationist idea of agents 'constituting' structures. This usage reinforces arguments about the intensity and direction of the causal links in models of the transformations effected by human activities in relation to urban landscapes and other features of the physical environments of societies. Further, it directs analytical attention to the observing subject's representations of environments and social and economic processes, and hence to empirical study of the ecological competence of agents. The image also suggests a sense of agents becoming, or 'constituting' in another sense, parts of the environments they create. This metaphorical device provides a useful check against excessive spread of the notion of the externality or 'separateness' of natural systems from human cultures. Further, invoking the structurationist parallel is a useful reminder of the importance of individual actions in relation to environments. Ecological phenomena are thus linked with the growing study of 'everyday life' in social theory and empirical sociological research (Crook, 1998).

Indeterminacies

Global environmental discourses contain turbulent mixes of conviction politics and arguments deferring to the epistemics of uncertainty. The two are not necessarily incompatible. Strong policy support for precautionary principles is based among other things on observations about the uncertain, or even self-defeating, consequences of actions that impact on natural ecological sytems. As Lorenz famously expressed the point, echoing observations on indeterminacy by Poincaré in the early 1900s (Rice, 1997: 94–5), a flap of butterfly wings in Brazil can lead to a tornado in Texas. However, suspicions that indeterminacies are ubiquitous in natural and social systems also create obstacles for the social-action and public-policy orientation of environmentalist arguments. There are also implications for theory projects. Four framings of the problem can be noted. These are respectively the arguments about uncertainties that arise from constructionist logics, from the scientific enterprise, from observations about ecological and also about social sytems, and from arguments about the nature of actions and their consequences.

Constructionist logics

The first of these arises from cultural critiques both of scientific epistemologies, and of the social and economic positioning of science-based knowledges in post-industrial societies. As was noted in chapter 4, the idea that environmental problems and scientific arguments are culturally constructed suggests a posture of relativity that sits uneasily with key assumptions and methodologies of environmentalism. If environmental problems do not 'objectively' exist, discussions of how best to solve them are either misguided, or, for the observing subject of them, just part of an interesting unfolding narrative. Environmentalist accounts tend to interpret the fact of the multiple representations of ecological phenomena as evidence of differing perspectives on an objective reality. The subjectivist critique, that is, is not taken to invalidate the hunch that core statements of environmental problems are true. Thus, though interpretations differ, the 'reality that refuses to go away is the world environmental crisis' (Anderson, 2000: 106).

The scientific enterprise

Secondly, and more importantly for policy-relevant environmental argument, uncertainties of a different kind are inherent in the traditional definition of the scientific project that underlies ecological and earth-systems research. As Pouyat has expressed it, 'The basic nature of science is that there will always be uncertainty. Science does not discover "truth"; it brings us closer to truth by attempting to falsify hypotheses – a process of elimination that reveals which possible answers are wrong' (1999: 282). Sociological processes of constituency-making in science can thus be protracted. This was the case with the emergence in the 1980s and 1990s of a research community on environmental endocrine disrupters and studies of their effects on wildlife populations and human health (Krimsky, 2000: 16–24). Data in some environmental-policy areas, such as on wildlife species, are notoriously difficult to collect; and since data-sets may be put together in different ways, there are often difficulties of inter-user translation and aggregation. Interpretations and findings are thus necessarily provisional. In science, it is the ongoing process that counts. Assumptions about the inherent complexities of dynamic systems reinforce the point (Hannon and Ruth, 1997). The ingrained sense of caution often assists environmentalist advocacy. It leads naturally to support for the precautionary principle in policy-making. However, it can also thwart the convictions of the impatient that certain actions need to be taken, or others prevented, sooner rather than later. Tensions result, too, within areas such as applied ecology, conservation biology and the study of geoenvironmental indicators, as researchers try to respond to demands for more expeditious, policy-relevant choices among conflicting interpretations.

However, uncertainty does not automatically lead to the conclusion that nothing is known about the global environment, or that scientific findings can never be a basis for action (VanAsselt and Rotmans, 1996). Nor does it equate scientific processes with those of other communities. Knox paints a warning picture of late eighteenth-century London discourses in which there was 'no space for the promotion of value-free knowledge'. The salons 'became the domain of fanciful poets and unhinged prophets; coffee-house wits were replaced by ribald and raucous rogues of the tavern' (1999: 451–2). A critical feature of environmental debates is thus the dynamic between complaints about the pretensions and consequences of the natural sciences, on the one hand, and alarms sounded by the practitioners and supporters of these about the failure of outsiders to appreciate the nuances of their art, on the other. Even when these are not made explicit, scientific statements come equipped with qualifications and reference to the bases of claims, or, as is the case more typically in the social sciences, they comprise propositions about the relations among variables expressed in probabilistic forms. Other regions of environmental discourse are not so restricted. These often lack in a sense the functional equivalents of the 'evidentials' – verbal suffixes indicating the source, reliability and other attributes of the information imparted and the communicator of it – found in some North American aboriginal languages. 'Clearly', Macauley adds, 'life would be much harder for politicians and advertisers' – and environmentalists too – 'if they were obliged always to provide an indication of the reliability of their claims' (1994: 156).

Some postmodern critiques draw on the uncertainty logics of various regions of the sciences, for example arguments in relation to the study of elementary particles about the constitutive effects of observations, as evidence of the inherent failures of their assumptions. Debates in quantum theory in physics can to some extent be seen as taking place between the contrasting views of realism – 'that there is an independently existing objective reality which we aim to describe in science' – and those that maintain that the scientist's categories for describing the world are 'predetermined by the categories of the understanding (and ultimately by the categories of perception) rather than developing out of the world as it is' (Krips, 1987: 127). But as Kosso points out, even the latter kind of approach does not necessarily support the conclusion that there are no properties of systems which are observable or knowable. Rather the argument points to complexity. 'Properties that we thought we understood and that we thought applied to things in simple, absolute, determinate ways, apply only in complicated, unexpected ways' (1998: 180–2). This way of looking at natural phenomena is echoed in many arguments about dynamic ecological systems.

Systems uncertainties

Thirdly, notions of uncertainty are inseparable, more specifically, from the study of ecological systems and processes (or, in a more restricted definition of the problem, from certain methodological approaches to the understanding of these). The history of the modelling of processes such as the population growth of an organism in theoretical or mathematical ecology has highlighted these constraints. In air-quality modelling, the science continually improves but the predictive power and other measures of model performance often tend to lag (Russell and Dennis, 2000: 2284). The issue has had profound implications for climate-change debates. The predictions for weather in western Canada in the summer of 1999 were for lower precipitation and warmer-than-average temperatures. Instead the summer was exceptionally wet and cold, and even included snowfalls, with severe consequences for farmers (McCarthy, 2000: 2).

The inference is not necessarily that improvements of data and modelling capabilities can eventually remove such difficulties, though that may in practice be an assumption driving research. Rather, the argument is based on assumptions about the nature of complex systems generally, and on the inherent problems of understanding these. As Gollub and Cross have written, 'Many aspects of nature are essentially unpredictable over the long term, even when quantum effects are completely absent' (2000: 710). Wagner suggests that as a result, the language of causality that has traditionally characterized many interpretations in the biological sciences has decreasing applicability. The problems, he writes, have become in a sense more analogous to those of economics and the social sciences generally. The apparent certainties of causal logics are in effect undermined by studies of non-linear systems, and research relating to chaotic dynamic systems generally (1999: 84–6).

Ecosystem dynamics, Pahl-Wostl argues, have an 'unruly nature', in which there is little place for 'traditional ideas of clear interaction pathways and defined cause–effect relationships' (1995: 192).

A bigger set of uncertainty problems results when study of ecosystem dynamics is linked with investigation of social and economic processes. Such complex links are central to research on natural resource use, health, toxic chemicals and many other environmental-policy issues. Changes in traditional diet and lifestyles among native Canadians, for example the Cree of northern Quebec, have resulted in a high incidence of diabetes and other health problems. The organization of fisheries has historically shaped and in turn reflected the social structures of coastal communities. Consideration of quotas and fisheries data is inseparable from these larger contexts, as in the disputes which broke out in Spain in the late 1990s as women's groups disputed the traditional restriction of fishing rights to males. Notions of complexity and uncertainty have played a growing role in research in the social sciences. The issues, as for the natural sciences, are different from those of incomplete knowledge. The condition of uncertainty stems from what Faucheux and Froger, in their discussion of environmental decision-making, suggest are inherent indeterminacies of the processes involved. Indeterminacy, they argue, 'should be considered as a significant feature of social life and action' (1995: 33). Environmental-policy debates characteristically centre on complex problems. In such settings, the rates of change of variables depend 'in nonlinear fashion on the variables themselves' (Gollub and Cross, 2000: 710). Since we cannot have full knowledge of initial system states – to return to the butterfly-wings example – predictive capabilities are bound to be faulty (Rice, 1997: 94–5).

This kind of concern leads Pahl-Wostl to criticize the applicability of the more conventional linear dynamic reasoning of ecology to study of the complex processes of climate change. An initial depiction of this problem, she points out, requires clarification and quantification of the multiple links in complex causal chains like society–GHGs–climate–ecosystems–society (1995: 196–200). The difficulties are reinforced by imperfect information about temperatures and other changes, and also by disputes over techniques for collecting climatic data (for example, as between terrestrial and satellite-based sources, or more specifically over the biases built into land-based data from monitoring stations near to cities) and over rival models of change processes (Houghton et al., 1996).

In such settings the views of social and economic actors themselves shape analyses. The catastrophic collapse of northern cod stocks off eastern Canada in the 1980s represented in part a failure of data and scientific modelling, but it also revealed the influence of a wide variety of interests and community concerns on scientific assessments and fisheries management policies (Finlayson, 1994). Conflicting views persist on the highly politicized issue of the amount of agricultural damage done by elephants in East Africa. Data from some studies are used in critiques of conservation programmes, while other studies indicate this damage is seasonal and that saplings can compensate at other times (Kabigumila, 1993).

General environmental accounts come equipped with their own descriptions of the causal links thought to be at work in such processes. Noorduyn and De Groot have summarized the 'generalized causal story of environmental science' in the following terms:

> Certain economic, social and cultural patterns in society, together with physical conditions, invite actors to use their environments in a specific way. This can cause environmental damage. The same patterns invite the actors to roll off the negative environmental consequence of their acts onto other individual actors, collective system levels, nature, or future generations, thereby influencing the environmental basis of these categories and sometimes changing the societal patterns. The environmental damage causes new activities, either lessening the negative effects or leading to more damage, depending on the societal and physical possibilities of the actors. This causal story ends when the environmental impacts reach normatively relevant victims, defined in the general goals of environmental policies, in terms of human health, economic sustainability, the protection of biodiversity, and so on.
>
> (1999: 35)

Built into this story, they suggest, are beliefs about its explanatory and prescriptive power, and the capabilities of those skilled in telling and expanding on it. Thus environmental science claims to be able to 'integrate knowledge from all elements of such causal chains to design curative or preventive policies for sustainability.'

Consequence appraisal

Fourthly, considerations of uncertainty arise in the context of actions, whether of individuals, NGOs, governments or others. Actions imply sets of expectations on the part of actors concerning anticipated outcomes. These do not neatly coincide with actual results. Studies from social psychology suggest that actors deploy often ingenious strategies to handle evidence of conflict between the two in post-decisional evaluations. Problems of this kind are widespread in environmental discourse and action.

Pursuit of one set of goals can thus hinder the chances of success of another. As noted in chapter 6, this applies within the broad range of environmental policy objectives, for example in the unforeseen effects of expanded alternative-energy projects on wildlife species. In parts of Thailand some forest-protection programmes in the 1990s led to problems for wildlife conservation projects because of reduced demand for the labour of elephants. Indian political debates in the early 2000s over the reallocation of parliamentary seats prompted criticisms that existing rules created disincentives for population policies, and unfairly discriminated against states with successful birth-control programmes.

Interference in ecological systems generally produces unanticipated results, or new sets of problems. There were widespread and unpredictable regional ecological effects of the forest-burning for land-clearing and agricultural development in Sumatra and Borneo in the late 1990s. Deforestation, for whatever reasons, exacerbates flood-management and other problems. It was a factor worsening the destructive effects in Central America of hurricane Mitch in 1998. Chinese policies in the 1950s aimed at the eradication of small birds in order to protect crops were reversed in the late 1990s because of concern over their ecological consequences. Soviet large-scale irrigation projects have had persisting ecological effects such as the destruction of much of the Aral Sea. Debates in the early 2000s among groups in Bangladesh about options to control floods have focused among other things on the possibly self-defeating, long-term ecological consequences of large-scale, capital-intensive control projects. The development by organisms of resistance has traditionally been a chronic problem of pesticide use. In the late 1990s there was early evidence of possible pest resistance to genetically modified crops, which are designed in part to include pest-control mechanisms.

Some of these problems do not involve 'unexpected' consequences. They reflect rather the persistence of sociologically embedded, and ecologically significant, cause-and-effect beliefs on the part of different actors. The growth and spread of scientific knowledge of natural systems, and of their interconnections with cultures and economies, is thus a foundation for environmental governance and, more specifically, 'a key resource in adapting to and managing extreme environments' (Kroll-Smith et al., 1997: 13).

Uncertainty problems, and the interpretive responses associated with them, are also features of transnational and transcultural interactions. There were catastrophic consequences in terms of the spread of diseases of nineteenth-century European colonialism. Although these were not anticipated, their gradual appearance gave rise to multiple rival accounts of causes and effects, and of the viability of different control strategies (Bewell, 1999: 27–65). More generally, Jervis's discussion of system effects in international relations (1997) develops the theme of multiple and unanticipated consequences of actions at the contemporary interstate level. These effects can generate resilient arguments based on conflicting interpretations of action–consequence linkages. Western liberal and conservative cold-war critics each criticized the allegedly self-fulfilling nature or paradoxical consequences of the actions advocated or practised by the other. There were similar discourse dynamics in the exchanges between the different coalitions of opponents and defenders of Western military action in relation to Kosovo in 1999.

Sociological models provide a diversity of approaches to uncertainty phenomena of these kinds. While Giddens and others tend to emphasize the knowledge, skill and valid expectations that individuals bring to bear on problems in their daily lives, other sociologically based models of action emphasize uncertainties. These are often more relevant to study of environmental problems. Portes (2000: 7–8) develops a typology of five sets of connections between expected and actual outcomes. These are when (1) apparent or stated goals are not the real ones; (2) the goal achieved is not the outcome intended; (3) objectives arise in the process of actions rather than as a prelude to these; (4) the result is unexpected in the sense that it is contrary to the original aims; and (5) stated goals are achieved not through planned actions but through fortuitous circumstances. The typology does not, however, specify the kinds of mechanisms that produce these varied effects. A fuller account of the processes would have to incorporate study of both interactions among actors, and of the effects of structural variables. Studies of environmental actions have identified similar sequences. The third, for example – where objectives are defined as a result of, rather than as precursors to, actions – recurs in what Voss terms 'ill-structured domains'. These, he argues, are prevalent in many areas of social policy and to some extent in economic policy. In such policy areas objectives tend to be loosely defined, and approaches and strategies generate sets of constraints not apparent at the outset. More specifically, it is the process of problem-solving which in turn drives discourses of reasoning about issues and objectives (1988: 74–5, 82–3).

Glocalities

These points bring us to consideration of the environmental effects of factors that check globalization processes, including the activities of agents designed to counter these. While some critics of globalization draw in part on evidence of its environmental costs, it cannot be assumed that the ecological consequences of activities designed to counter globalization are necessarily beneficial. Further, some such checks are 'natural,' in that they operate separately from agents' perceptions of globalization or constructions of their own actions as elements in counter-processes. Thus in models derived from classical IR, processes of economic globalization can take shape, and influence political and economic environments, only within limits prescribed ultimately by states. Frameworks stressing ethnicity or religious beliefs identify different factors in global society, but likewise see these as limiting factors. The focus in this section is on the environmental significance of the responses of states and civil societies to globalization.

States and globalization

Much environmental discourse focuses on problems arising from the diminished autonomy and capacities of states under conditions of economic globalization. The issue is part of a broader critique of the failures of the contemporary state. As a result of FDI, trade-regime and other external pressures, and the linkages of these with internal economic changes, states have gradually lost some, perhaps a significant part, of their capacity to deliver public goods. The neoliberal rationale from the mid-1980s has been that interventionist states, top-heavy with regulatory power and weighed down with debts and deficits, are inefficient, poor managers, and inadequate providers of services to their citizens. Thus the 'residual state', in Cerny's view, 'faces crises of both organizational efficiency and institutional legitimacy' (1996: 598).

The enfeebling process is linked in part to changes in international institutional arrangements. Some of these are analytically separable from globalization effects, though the two have become interconnected. They occur particularly among OECD countries. One dynamic has been the 'internationalization' of public policy, defined as 'a process that starts when at least one aspect of domestic policy begins to depend on or be affected by forces beyond the borders of the state' (Doern et al., 1996: 3–4). In relation to environmental policies, mechanisms of change include the framing of definitions of problems in discourses, particularly the standard environmentalist focus on the transnational character of pollution and conservation issues (Economy and Schreurs, 1997: 5). More generally, regime changes result, first, from autonomous responses to problems in environmental, health and other public-policy areas, and to pressures from civil-society groups for expanded IGO capabilities. A second source of change, discussed in chapter 7, includes pressures for the harmonized and IGO-monitored regulation by states of activities that might have trade-distortion effects. Environmental regulations, for example, can be used as means to erect non-tariff barriers. Whereas the first set of effects potentially protects a significant part of the regulatory reach of the state, for example by facilitating the creation of interstate regimes based on large geographical regions, the latter defines intrinsic limits to state power.

However, states have clearly not become redundant. They are critical institutions in strategies of adjustments to market failure, and in proventive standard-setting and monitoring in areas such as urban water-quality. They have crucial roles in relation to natural disasters like the earthquakes in Armenia in 1988 and Kobe in 1995 through such means as the design and enforcement of high-rise-apartment and other building standards and emergency-measures preparedness. State capacity-building is also required to tackle phenomena such as the growing involvement of organized crime in the transborder hazardous-waste trade in North America from the late 1990s. Climate change is among several sets of global issues that require policy responses premised on long-term state intervention at critical junctures of complex social- and economic-ecological processes (Le Prestre, 1997: 75–6). States remain central to the agendas of environmental NGOs through their regulatory authority and power to effect sanctions against pollution-code violators. They are also indispensable associates of firms as providers of consular services in foreign markets and sources of a broad-ranging fabric of regulatory and monetary policies, and through activities designed to strengthen intellectual-property regimes (Weiss, 1998). States also remain the primary actors in international environmental law. Multilateral environmental agreements, the primary vehicle for much transboundary environmental cooperation and dispute-resolution, are embedded in state contexts. They are effective to the degree that states have the political will to ensure this.

On the other side of the equation, some of the actions of states are also detrimental to civil societies and environments. I will come back to these problems later.

Globalization in these contexts does not incrementally eradicate states. Rather it 'transforms the ways that the basic rules of the game work in politics and international relations and alters the increasingly complex payoff matrices faced by actors in rationally evaluating their options' (Cerny, 1995: 596). The changing rules of the game that have a bearing on environmental governance reveal significant differences among the advanced industrialized and developing economies.

In the societies of the North, environmentalism both in and outside governments is structured in economic settings governed by the momentum towards enhanced trade liberalization and capital mobility. Environmental questions are affected at multiple points in the agendas of the reduction or elimination of NTBs and subsidies, the protection of intellectual property rights, government reorganization, and deficit-reduction and debt management. More assertive forms of regulatory environmentalism have been subdued. The goals of environmental protection and sustainable development are pursued in terms of more specific objectives such as the greening of government operations, the professionalization of environmental-policy expertise and its spread among public and private sectors, adherence to the polluter-pays principle, indirect encouragement to the growth of environmental-technology industries, and attention to market-based environment-policy instruments such as tradable pollution permits. Globalization has also created sporadic political alliances between environmental and labour organizations. Further, the range and definition of issues conventionally counted as environmental differs from those of the South.

Environmental thinking has thus become entrenched within OECD governments, including those of sub-national jurisdictions. But its influence is variable and contested. Environmental agencies are typically limited in their capacity to affect the strategic directions of Western governments. They face increasingly tough bureaucratic battles in protecting the national environmental-policy *acquis*. Agencies may be less well equipped as a result to resist pressures for access to national parks from mining or resource-exploitation firms, or to counter other threats to the ecological integrity of protected areas. In federal systems, national governments are only part of a complex governance mix in which state and provincial governments are key players; and the environmental- and resource-policy approaches of the latter are increasingly shaped by incentives and constraints from the global economy. Japan's environmental agency has traditionally been weakened by prevailing economic-growth goals, as well as by its location outside the traditional departmental structures represented in the Cabinet. It also lacks a constituency of effective national environmental NGOs of the kind present in other OECD countries.

Environmentalism nestles in different economic contexts in the South, and also to some extent in the transitional economies. In Indonesia, government capacities to formulate and implement environmental policies have been formed in association with development-assistance funding from IFIs and Northern states (Boardman and Shaw, 1995). However, capacity-building has been constrained by the wider contexts of government organization, democratization, definitions of economic-development priorities, and the incentives conditioned by global economic factors that often tilt choices of development strategies towards those with significant ecological costs. The World Bank argued in relation to Indonesia in the early and mid-1990s that state intervention was required in some areas of environmental and resource policy, for example for ecosystem protection, but reliance on this instrument, as opposed to moves towards the market pricing of natural resources, risked contributing to problems such as urban pollution and over-use of groundwater (1994: 26, 210).

Inadequate environmental protection in Southern economies results in part from beliefs that environmentalism is more a brake on development than a strategy for promoting economic sustainability, or that it is a tool that primarily serves Northern state and private-sector interests. Hecht has argued that 'although all policy makers can see the impact of environmental degradation, especially among poorer urban communities, policy makers in many developing countries are often unconvinced that environmental protection does not hinder their economic growth' (1999: 113). Systemic factors associated with globalization, for example the effects from short-term flows of finance capital (Bhaduri, 1998: 153–4), remain significant contexts of environmental policy-making by redefining and constraining the scope for change in national economic policies and actions by local communities.

Alternative long-term approaches to change, as highlighted for example in the 2000 Havana declaration on the New Global Human Order, include removal of limitations of trade and labour access to Northern markets, and greater provisions for technology transfer. Not all such steps, however, are likely to lead to environmental improvements. Specific constraints relevant to environmental policies in the South include information-technology gaps, the limited availability of scientific and professional expertise in key environmental-policy areas, the costs of effective representation through large and adequately serviced delegations in multilateral conferences, and chronic shortfalls in environment-agency budgets.

There are some structural similarities with conditions in the transitional economies of Eastern Europe and post-Soviet Central Asia. The circumstances of these are as diverse as those of Southern societies. As Sölderholm has argued in the context of Russian environmental problems, however, major institutional change is a requirement for effective environmental-protection and pollution-control measures by governments. Without this, Western policy ideas of market-based environmental policy instruments have little relevance (1999: 403–4).

There was increasing support generally during the 1990s for such instruments, for example as a basis of international climate-change arrangements. A related development was debate sparked by calls for the wider and more formalized participation of firms in the making of international environmental policy. The argument gained ground as a result of anticipation of and reactions to the 1997 Kyoto climate-change conference. There was growing MNC resistance, especially in oil and gas sectors, to the prospect of deepening regulatory interference by governments embroiled in the politics of compliance with international regimes. Observers sympathetic to market models increasingly argued that extensions of command-and-control environmentalism to international levels were in any case bound to fail. Companies, by contrast, are seen from this perspective to have an interest in stabilizing economic environments, in responding to shifts of consumer attitudes, and preparing economic sectors for future generations of consumers. Dickinson estimates that in view of the political constraints on governments in enforcing post-Kyoto emissions controls, effects on predicted global temperature rises in the period to 2100 may be less than 1°C. Moving away from regulation, he argues, and using the 'carrot of energy savings and clean energy', would be a more efficacious route to control (2000: 2). More generally, Daily and Walker have argued against efforts to strengthen government capacities in the area of environmental policy. Promoting instead lead roles for firms and industry associations, they argue, is a better route to sustainability. It leads to stronger economies and hence to changing industrial practices, producing 'cascading effects through all the determinants of environmental impacts'. They acknowledge that the reasoning is contrary to the general direction of much environmental discourse, but maintain that 'failing to engage industry may well ensure that sustainability remains largely an academic debate' (2000: 243–5).

Partnerships of this kind, however, are unlikely to be balanced ones given the economic power of companies, and the multiple pathways available to them for influencing governments. This is especially true of MNC relations with governments in the South. Enlightened self-interest on the part of firms normally prompts some form of commitment to environmental goals, if only to ensure greater long-term stability in markets or to ward off consumer or shareholder revolts in home countries. Self-regulatory practices, for example in the chemicals industry, have become more widespread in the North. However, it is unclear how good practices could be evenly spread among competing and differentiated economic agents in the absence of proactive roles for governments and restructured IGOs. Lack of regulatory and scientific capabilities on the part of governments in the South have led to problems from the use and handling of agricultural chemicals, for example, and the driftnet-patenting of biological organisms by foreign companies. Withdrawals from modestly interventionist roles have also had significant consequences in terms of civil strife. There was deepening social conflict in Bolivia in the late 1990s, for example, following the government's IFI-encouraged water-supply privatization, until this step was reversed in 2000.

Round-table and other inter-stakeholder formats have nonetheless taken increasing hold at local levels in Western states. In relation to endangered-species protection, diverse institutional settings have emerged in OECD states among nature-conservation and other environmental NGOs, government agencies at multiple levels, and corporate and household landowners. These suggest lessons for emerging transnational governance practices. However, they are also prone to conflicts and breakdowns. Environmentalist advocacy can be weakened where it entails neglect or casual dismissal of the declaratory environmentalism of MNCs, impatience with policy forums that encourage corporate participation, faith in exclusively regulatory policy instruments for pursuing environmental objectives, a resort to scapegoating, or a reluctance to discriminate critically among the motives of private-sector players. Governance in many environmental policy areas requires more institutionalized formats of discourses and collaboration among multiple actors.

Civil societies and globalization

So far we have been looking at the environmental implications of globalization processes entailing change in the capacities of states. Impacts on societies and cultures, and the responses of these, also affect environments. Some of the critical effects are mediated through government actions. However, accounts of ecological costs developed in critiques of globalization effects are sometimes grounded in practice in other criteria. Many reflect values relating primarily to social justice, equality or inter-ethnic peace, rather than to ecological integrity. There are accordingly divergent assessments of cause-and-effect relations among factors.

Some globalization effects work indirectly to strengthen civil-society actors and their roles in relation to environmental policy and programmes. As state authorities in Northern states have edged out of innovative or leadership roles in areas of social and environmental policy, niches have been created that facilitate expanded roles for NGOs. Government environmental agencies have a stake in promoting diverse forms of partnership and joint-venture arrangements with non-governmental groups. Many of these have taken on attributes of quasi-governmental organs, in terms of both policy development and programme implementation. Roles vary considerably among OECD countries. In Japan, the creation of a wide range of support and self-help groups in social-policy areas in the late 1990s followed widespread criticism of the government's emergency-measures failings immediately after the 1995 Kobe earthquake. Significant changes have been observed in relation to welfare policies in the UK. Groups organized around issues of gender relations, race, age and disability, and a variety of self-help groups, have provided direct non-state-based services, for example for refugees and persons with AIDS. At the same time these groups have increasingly shaped social and political discourses, the agendas and consideration of policy options by governments, and their practices of service deliveries (Williams, 1999: 668–9).

Beck has been among the more optimistic observers of globalization. Rather than making individuals subject to new and more powerful transnational economic forces over which they have little control, globalization, he argues, is instead helping to structure a 'second age' of modernity characterized by respect for human rights in an overarching cosmopolitan framework (2000: 80), with 'globality at the heart of political imagination, action and organization' (1998: 28–30). Effects are varied, though, as pressures are translated through the structures of national societies. Differing patterns of ethnicity in Western states may affect their respective responses and capacities in relation to economic globalization. A resurgence of identity politics has been a related consequence. In Japan, the 'almost obsessive' preoccupation with questions of self-definition, central since the nineteenth-century opening up to the West, has intensified since the 1940s; it has found expression, for example, in the protracted debates on the kimono (Goldstein-Gidoni, 1999).

In the South during the 1980s and early 1990s, the democratization effects of globalization were less commonly observed than those of homogenization and constraint. Civil-society actors lacked the capabilities to carry out extensive social and quasi-governmental tasks. As a result,

Kothari wrote in the early 1990s, there was 'little scope for alternatives'. Globalization meant 'a reduction of the richness and complexity of civil society to a marketplace, turning a variety of civil societies into parts of one huge global supermarket' (1993: 119–20). This critique applied especially where, as in Indonesia in the 1970s and 1980s, development was 'engraved on people's minds [as] basically the concern of the government ... with the role of the community being reduced to "participation" as recipients or passive observers' (Johnson, 1990: 78).

However, spreading networks of local groups organized around issues such as air and water pollution were a more characteristic feature from the mid-1990s, particularly in the NICs of Southeast Asia and Latin America. These changes have been associated with diverse societal and cultural responses to global and regional economic change. Studies of globalization in its more complex forms emphasize the enhancement of difference within and across borders, particularly in the characteristically post-colonial 'murky realm of dislocated and displaced identities' (Fludernik, 1999: 30). Thus in the 1990s and early 2000s globalization effects have been inseparable from internal democratization processes. In Indonesia, political and social-policy ideas were transmitted in the 1990s in many ways. Local community debates on gender, tourism, environmental problems, international politics and a variety of other topics arose in 'resistant readings' of religious texts in prayer groups that contested the views of state authorities (Weix, 1998: 409–10). The responses and strategies of groups in societies of the South to the changing rules of transnational games thus tend to be more complex and proactive than earlier criticisms anticipated. Adams, also writing on Indonesia, has criticized the notion that expanded tourism leads through museumification to passivity and a loss of agency. In Tana Toraja, community groups have developed 'ingenious strategies' to protect traditional identities in the face of increased contact with foreign tourists, including self-conscious redefinitions of history (1997: 310, 318).

Democratization and identity responses such as these have varied environmental consequences. Civil-society organizations themselves have widely differing views on environmental questions and the priority they attach to these. Responses also vary to the government policies conditioned in part by globalization. For example, environmental damage has resulted from expanding human settlements in officially designated greenbelt areas around Mexico City. The movements were prompted initially by economic change, and social conflict was exacerbated by government responses. Government policies have led directly or indirectly in many countries to internal movements of peoples. These have diverse ecological consequences, for example in the mounting pressures on forest ecosystems in Brazil or, as in Thailand, on local agricultural economies. Especially where such movements take place across borders, ecological factors interact with others in complex sets of globalization effects that include armed conflicts. The causal sequences associated with modernization theory traditionally tend to link conflict to poverty, and to see liberalizing trade and state deregulation as ameliorating conditions. Critical accounts emphasize instead the systemic economic factors that reinforce inequalities and resource pressures, and hence precipitate conflict.

Ecological factors can be identified as causes or effects in some conflicts. Findings vary from case to case, however, and are not necessarily indicative of a general causal relationship between environmental stress and violent conflict. Water resources have been complicating or reinforcing factors in interstate conflicts, as in the Middle East, but have not directly produced these (Lowi, 1999: 393). Wars have multiple ecological effects. Civil strife in Angola in the 1980s and 1990s destroyed wildlife habitat. Among other factors, energy shortages reinforced extensive deforestation, which itself has had far-reaching ecological consequences. The NATO bombing of a Serbian petrochemical complex on the Danube had serious environmental impacts in the 1999 war. The earlier civil war in Bosnia led to widespread deforestation and soil erosion around cities because of fuel shortages; however, for these reasons it also led to significant improvements in air quality in cities (EEA, 1998: 283).

Global economic factors, then, have consequences for ecological change since they help define the broader setting of constraints and incentives within which governments operate. Economic change and social conflicts increase flows of environmental refugees and precipitate other movements of peoples. The Brazilian government has opened up large areas of land for private ownership, and adopted programmes to speed up settlement in distant frontier regions. This has led to an increased incidence of violent conflicts among landowners, squatters and indigenous groups. Significant environmental consequences have resulted from accelerated deforestation (Alston et al., 2000). Noorduyn and De Groot review several recent cases from Africa. Each involves movements of people within countries as a result of incentives created by government programmes, and complex interactions among social, economic and ecological factors. In Mali, conflicts and serious ecological problems have resulted from overgrazing by cattle, which was a consequence of government plans to develop drylands. Government encouragement to charcoal-making from trees in a community in Somalia similarly had serious environmental consequences. Developments were associated with significant social unrest; and violent conflict was prevented only by the threat of military action from the government. The official promotion of rice-growing in Ivory Coast has also led indirectly to extensive deforestation. In this case changing economic incentives meant that the traditional sustainable crops of yam and cocoa were no longer competitive (1999: 30–1).

Globalization effects are also evident in the way issues on traditional environmental agendas are redefined. There have been growing arguments in CITES, from both producer states and some environmental groups, in favour of the sustainable use and marketing of endangered wildlife species rather than continued reliance on traditional protection measures. At the 2000 meeting of parties in Nairobi, Cuba unsuccessfully proposed legitimizing sustainable trade in the hawksbill turtle, arguing that apart from economic benefits this would be a better route to species conservation. A similar case, based on increased economic benefit to poor rural areas and more effective protection of wildlife populations, has been made for the revitalized elephant and traditional big-game hunting economy in Zimbabwe.

Critical ecological theory

Five main bases of global environmental enquiry were discussed in previous chapters. These were the broad approaches characterized, respectively, as those of the geosciences, constructions, ethics, individualism and governance. It is possible to devise multiple sets of theoretical linkages among the diverse elements in each of these. This chapter has suggested that interdisciplinary perspectives drawing respectively on ecological and economic reasoning, each broadly defined, present combinations of divergent and complementary insights about global environmental problems and their significance.

The divergent tendencies between these stem partly from the greater affinity of many ecologically grounded narratives with scientific epistemologies, and hence with arguments about the uncertainties or indeterminacies of dynamic systems. Consideration of ecological processes is more muted, if present at all, in globalization perspectives, even those that elaborate on environmental problems. Other strands of ecologically based approaches connect with normative discourses, for example on obligations to nature, future generations, or wildlife species. These topics are generally absent in ethical debates based in political economy, which deal more unequivocally with considerations of justice arising out of human needs and inequalities, within and between national societies. Approaches to globalization also tend to stress the transformative character of contemporary economic change. While there is emphasis on transformation in geoenvironmental studies of the effects of human activities on natural systems over time, adaptation metaphors in ecology have traditionally suggested alternate, and increasingly less realistic images of societal responses to relatively static environments as contexts for enquiry.

These two perspectives on critical ecological theory are not incompatible. The differences between them are more tendencies than structural impediments to mutual influences. They point respectively towards either general-theory building efforts *grounded* in the ecological and earth sciences, enriched with insights from the social sciences; or to ecologically *enhanced* frameworks of international political economy and global society.

9
Silent String

The study of global society opens up spaces beyond those inhabited by states. Globality points to civil-society groups and individuals, and to ethnicities and identities other than those associated with nations; to images and representations of social worlds, and the normative discourses of actors on these; to markets and their interrelations with cultures; and to differing conceptions of governance. Natural systems interact in complex ways with societies and economies, and with the globalization factors transforming these. Understanding environments requires investigation of the forces of economic and cultural change in world society, and it leads back in turn to questions that capture a richer sense of the character of ecological processes.

Environmental change took on an increasingly global character in the late twentieth century. Instances of pollution were formerly local. These expanded in scale and were redefined in global contexts as contributions to changes in atmosphere-climate and other complex earth cycles. Multiple human activities have had deepening effects on the biosphere, with wide-reaching implications for economies and societies. An increasingly urbanized Earth recalls the image of the city-planet Coruscant at the start of the *Star Wars* cycle. Interlocking processes of ecological and economic change have thus become transformative. Geoenvironmental studies are identifying the accumulated changes effected by human activities over the varying time-scales of recent decades, the last few hundred years, and the last several thousand years.

Systemic change has highlighted the normative character of many environmental problems. These touch closely on traditional issues of justice, as well as newer ones of right ecological behaviour. While nature and its resources can arguably be 'managed' for the general good, whether national or global, and while there is a historical record of environmental progress in the way Northern societies have handled at least some pollution problems, many issues remain both unresolved and also profoundly political and ethical in character. Some economic activities, by companies, states or households, are potentially self-destructive, harm human health, or reduce the quality of life for persons. Further, issues and actions are associated with uncertainties, a problem underscored in the scientific study of dynamic systems. This too implies the unavoidability of choices, and hence of politics. The varied discourses of sustainability, deep ecology and risk warn in different ways against the hazards of unchecked meddling with natural systems. In part, too, the ethical character of the issues stems from questions of scale. While phenomena such as habitat destruction or over-exploitation of fisheries are not novel, or even features of modernity, the transformative impacts of biotechnology, energy-consumption patterns and natural-resource use have increasingly given these a global, long-term character.

Global environmental change also raises old questions about the territorially fragmented nature of world society. Environmentalist logics often imply a cosmopolitan conception of political community. Such end-states are probably unattainable, but discourses on them, like those about obligations to future generations, illuminate current dilemmas of social action and governance. Institutional arrangements grounded in the policy prescriptions some environmentalist critiques draw from the facts of ecological interconnectedness – such as greater authority for NGOs – can be achieved only in limited fashion in a world of states. Governance issues often turn instead on the messier worlds of the politics of inertia and interest-based conflict. They lack the grandeur of visionary approaches to global environmental problem-solving. Despite the late twentieth-century growth of international environmental law, global institutions focused on environmental protection and ecological integrity remain conspicuously weak compared with those of many other governance areas.

Questions of global ecological change are investigated in multiple disciplines. These have protective dynamics that maintain boundaries. Such mechanisms promote creativity; without them, moreover, definitions of interdisciplinarity would collapse. Yet understanding the environmental domain also requires sustained transdisciplinary activities. Disciplines central to global ecological enquiry vary in their internal knowledge-arrangements and their flows of cross-border contacts. Paul suggests that in anthropology boundary-crossing has become a 'thing of the past', as the discipline has instead moved to 'purify itself' by dividing into largely non-communicating sub-approaches (2000: 9). The institutionalist tradition in economics, though, continues to seek out insights from sociology on questions such as the role of habit in decision-making (Sugiura, 1999: 260). The spread of critical-theory discourses in political economy and the social sciences generally has enhanced communications and cross-fertilization by facilitating the creation of metasets of research questions. Further, studies of both globalization and environmental change have increasingly encouraged a sceptical posture towards the traditionally national-society bias of specific social sciences.

Global environmental enquiry is none the less grounded in large measure in the natural sciences. This premise remains contested, however, even allowing for differing epistemological and ontological perspectives among scientific disciplines. To some, it appears to support notions of decision-making by experts, in national societies or transnational institutions, that are incompatible with democratic criteria. Scientific advocacy may also entail a spread of its methodologies into regions that cannot sustain them. Wilson argues that the problems of sociology and its allies will only be clarified and resolved when these disciplines accept the standard epistemologies of the biological sciences (1998: 186–92). An environmentalist-action perspective sometimes finds scientific conclusions excessively cautious and provisional, and failing even to justify preventive actions. Conversely, where science does support the precautionary principle it provokes criticisms, from more conservative or neoliberal perspectives, that all technological changes involve risk, and that risk is an escapable feature of economic and social progress. Science is also vulnerable to those aspects of environmentalism that reflect an anti-materialist posture, or that associate the sciences with damaging technologies, ethically suspect forms of rationality, and epistemologies that claim for themselves special status among cultural discourses. The collapse of science's 'ancient privilege' (Pels, 1995: 79) is welcomed triumphally by some critics. Horton, however, regrets the 'profound displacement in the position of science in our culture' (quoted in Grove, 1999: 391).

A reluctance to engage with scientific argument, as opposed to merely responding to the sociological fact of scientific activity, weakens social-ecological and other forms of critical enquiry. Significant policy and theory-building questions hinge on quantification of the links in general circulation and other ecometric models, research on geoenvironmental indicators of climate-change and other processes across time, and knowledge of the complex linkages between micro- and planetary-level processes. To use an old metaphor, we need more knowledge of the sundry threats to the string held in medieval and Renaissance accounts to signify the properties of the harmonic universe, and whose 'untuning' was contemplated by the fearful Ulysses in *Troilus and Cressida*.

Yet many of the questions touched on in environmental enquiry are also, as noted in chapter 2, those of the social sciences. They involve questions about how and why economies are organized in such ways that GHGs are produced, the factors that influence environmental decision-making by individuals in their daily lives, the internal workings of governments that shape their actions, and the significance for environmental problem-solving of the way global society is organized. Such questions have often been treated in the natural sciences either tangentially or as matters of common sense. Diversity in the social sciences, moreover, points to multiple directions for environmental enquiry. Wallerstein describes the social sciences in general as being caught in a 'pincer movement' consisting of forces from the 'sciences of complexity' on the one hand, and the pressure of 'cultural studies' from the humanities on the other (2000: 30–1). There are complementarities between the approaches of the natural sciences and parts of the social sciences to environmental questions. To some extent these affinities can be detected in studies of literature, for example by treating novels as producers of empirical and normative propositions about societies and cultures (Swirski, 2000). However, while some boundaries among the humanities and social sciences are porous, others are vigorously defended, particularly as social scientists have had to respond to diverse critiques that subvert traditional assumptions about observers, facts and knowledge processes. Investigations of the particularities of everyday life, and of topics in environmental ethics that ground these in considerations of moments of time and place, are among interdisciplinary routes to ecological enquiry influenced by spatial notions of globality.

Environmental discourses can usefully be approached too as exercises in theory. This does not mean there has to be consensus on this term. Environmental study rests comfortably on divergent and even conflicting styles. Alongside empirical enquiry into the 'brute facts' are approaches reflecting on the best ways to tinker pragmatically with, or overhaul, social- and economic-ecological systems. Normative discourses coexist uneasily with theoretical critiques in which environmentalism is treated as an integral part of social and economic forces of reproduction. Theoretical enquiry is also an end in itself, whether or not it has empirical or policy relevance. Yet while issues of critical ecological theory are central to questions of identity, social change, media and culture, they have had a precarious history in analyses of these. Alertness to the multiple meanings of theory is required for study of the environmental domain.

Like irises and dogs, environmentalist forms of argument are diverse. A variety of insider and hermeneutic, and outsider and distantiated, perspectives is useful for understanding environmentalism as a cultural phenomenon. Like post-Impressionism and postmodernism, it is one associated with specific spaces and times. It bears the imprint of the North of the second half of the twentieth century. Problems of air and water pollution, as well as the politics of environmental groups, have become more prominent in many regions of the South. Despite a growing integration among environmentalist and development discourses, however, suspicions of sustainability arguments as brakes on economic development, and as products of the interests of Northern actors, have remained significant checks on the global spread of environmentalist ideas and practices.

Research has converged from several directions to highlight the importance of micro-level change and its links with systemic properties. As noted in chapter 3, the traditional micro-level focus of ecological research has increasingly been joined by macro-level investigations, and both have been redefined in earth-system frameworks as part of the study of planetary processes. Environmentalist discourses have traditionally emphasized, perhaps iconized, individuals as the makers of social change, while at the same time urging attention to the global economic and social contexts of environmental problems. Sociological studies of globalization have likewise highlighted connections with traditional foci on individuals and localities. However, attention to micro-level processes has often been missing in governance discourses and practices, despite the emphasis in sustainable-development arguments on the requirement of democratic and participatory criteria in IGOs. In addition, knowledge of individual and other forms of microecometric decision-making is increasingly important for assessing the likely compliance responses of actors to different policy options, or in investigating the factors that influence decisions affecting ecologically related civil strife. These questions are also closely connected with enquiry into the multiple representations of problems present in environmentalist discourses.

The importance of the governance dimension of environmentalism, discussed in chapter 7, can easily be exaggerated. It is unlikely that long-term solutions to complex ecological problems will be found to be located primarily in this area. The varied forms of governance at issue include those organized predominantly around states, or the institutions of civil society, or restructured global institutions. Connections with domestic regimes are also central. As part of a larger class of risk problems, environmental problems are characterized by a requirement for sustained intervention by state authorities and other social actors; often by a complex or inchoate definition, and formulations that change during the course of action or as a result of vocabulary politics; by uncertainties in assessments of the likely consequences of different response options, and disagreements over the criteria for action; and by relentless patterns of continuous politics at domestic and transnational levels. And as environmentalism has changed, so has the nature of the governance problems.

Environmentalism was typically identified in the 1970s and early 1980s with groups constituting a new social movement. This casting has changed since then. A more characteristic guise since the mid-1980s has been as an extension of the administrative or deregulating state, rather than as a mobilizing force for social change. Even so, processes of interaction are complex. Environmental NGOs generally have retained critical, agenda-setting and other roles while participating in governance activities. State agencies, scientific organizations, environmental NGOs and others share the role of the 'nightingale floorboard', the mechanism built into some Japanese castles that warned of impending danger. The broader environmental capabilities of Northern societies have accordingly been structurally weakened by continued state withdrawals from social- and environmental-policy areas. Environmental governance has been characterized by mixtures of incremental progress and setbacks. There has been growing convergence among governance forms at different levels, for example as plurality models and multi-stakeholder formats devised at local and national levels have spread among transnational contexts.

The task-based character of functionalist governance arrangements, redefined with greater emphasis on normative discourses among actors, has a particular resonance with environmentalist debates on governance issues. Yet they also have some unsettling implications for these: first, by advocating the participation of various often contending actors – states, NGOs, scientific groups and, though less unambiguously, companies – in deliberative processes; and, second, by undermining key features of the environmental project itself by de-ideologizing it, fragmenting it into discrete tasks, and allying these tasks with others not conventionally recognized as 'environmental'.

Differing notions of globality are typically deployed in environmental discourses. They refer variously to environmental problems with causes and effects across borders; to similarities among the stocks of environmental problems encountered in different countries; to core features of environmental issues, particularly the requirement of common modes of problem-solving resting on NGOs and epistemic communities; to problem areas such as climate change that have consequences for all national societies; or, as in study of the economic and other processes of globalization, to conceptions making tacit use of geographical imagery. The latter spatial sense has become increasingly central in the study of global society. It facilitates enquiry into questions such as the impact of political or territorial boundaries on the production of environmental problems and the routes to problem-solving. And it has a temporal dimension that connects to the environmentalist preoccupation with future generations, as well as to the historical (social, economic and geological) study of glocal ecological pasts.

References

Abell, P. (1996), 'Sociological Theory and Rational Choice Theory', in B. S. Turner, (ed.), *The Blackwell Companion to Social Theory* (Oxford: Blackwell) 252–77.

Adams, K. M. (1997), 'Ethnic Tourism and the Renegotiation of Tradition in Tana Toraja', *Ethnology*, **36:** 309–20.

Ahuja, A. (2000), 'Glad to be Gaia', *The Times*, 15 May, **2:** 8.

Albert, M. (1999), 'Observing World Politics: Luhmann's Systems Theory of Society and International Relations', *Millennium*, **28:** 239–65.

Albrow, M. (1997), *The Global Age* (Stanford University Press).

Alston, L. J., G. D. Libecap and B. Mueller (2000), 'Land Reform Policies, the Sources of Violent Conflict, and Implications for Deforestation in the Brazilian Amazon', *Journal of Environmental Economics and Management*, **39:** 162–88.

Anderson, E. N. (2000), 'On an Antiessential Political Ecology', *Current Anthropology*, **41:** 105–6.

Aswani, S. (1999), 'Common Property Models of Sea Tenure', *Human Ecology*, **27:** 417–54.

Atkinson, A. (1991), *Principles of Political Ecology* (London: Bellhaven Press).

Avio, K. L. (1999), 'Habermasian Ethics and Institutional Law and Economics', *Kyklos*, **52:** 511–35.

Azar, C. and J. Holmberg (1995), 'Defining the Generational Environmental Debt', *Ecological Economics*, **14:** 7–19.

Barker, C. and G. Dale (1998), 'Protest Waves in Western Europe: A Critique of "New Social Movement" Theory', *Critical Sociology*, **24:** 65–104.

Baron-Cohen, S. (1995), *Mindblindness: An Essay on Autism and Theory of Mind* (Cambridge, Mass.: MIT Press).

Barry, B. (1999), 'Statism and Nationalism: A Cosmopolitan Critique'. In I. Shapiro and L. Brilmayer (eds), *Global Justice* (New York: New York University Press), pp. 12–66.

Bate, J. (2000), *Song of the Earth* (London: Picador).

Baudot, B. S. (1999), 'Dimensions of the Population–Environment Equation'. In B. S. Baudot and W. R. Moomaw (eds), *People and their Planet: Searching for Balance* (London: Macmillan), pp. 1–28.

Bauman, Z. (2000), 'On Writing: On Writing Sociology', *Theory, Culture and Society*, **17:** 79–90.

Baumol, W. J. (2000), 'What Marshall Didn't Know', *Quarterly Journal of Economics*, **115:** 1–44.

Baumslag, D. (2000), 'The Role of Rhetoric in Ethical Argument', *Dialogue*, **39:** 129–40.

Beazley, K. (2001), 'Why Should We Protect Endangered Species?' In K. Beazley and R. Boardman (eds), *Politics of the Wild: Canada and Endangered Species* (Toronto: Oxford University Press).

Beck, U. (1998), 'The Cosmopolitan Manifesto', *New Statesman*, 20 March: 28–30.

——— (2000), 'The Cosmopolitan Perspective', *British Journal of Sociology*, **51:** 79–106.

Beder, S. (1996), 'Charging the Earth: The Promotion of Price-based Measures for Pollution Control', *Ecological Economics*, **16:** 51–62.

Bess, M. (2000), 'Greening the Mainstream', *Environmental History*, **5:** 6–26.

Best, S. and D. Kellner (1999), 'Debord, Cybersituations, and the Interactive Spectacle', *Substance*, **28:** 129–56.

Bewell, A. (1999), *Romanticism and Colonial Disease* (Baltimore: Johns Hopkins University Press).

Bhaduri, A. (1998), 'Implications of Globalization for Macroeconomic Theory and Policy in Developing Countries'. In D. Baker et al. (eds), *Globalization and Progressive Economic Policy* (Cambridge: Cambridge University Press), pp. 149–58.

Bilmes, J. (1986), *Discourse and Behavior* (New York: Plenum Press).

Boardman, R. (1997), 'Environmental Discourse and IR Theory', *Global Society*, **11:** 31–44.

———(1999), 'Practical Things, Constricted Interests: David Mitrany and the False Security of Ecofunctionalism'. In L. M. Ashworth and D. Long (eds), *New Perspectives on International Functionalism* (London: Macmillan), pp. 156–69.

———(2001), 'Risk Politics in Western States'. In K. Beazley and R. Boardman (eds), *Politics of the Wild: Canada and Endangered Species* (Toronto: Oxford University Press).

Boardman, R. and T. Shaw (1995), 'Protecting the Environment in Indonesia'. In O. P. Dwivedi and D. K. Vajpeyi (eds), *Environmental Policies in the Third World* (Westport, Conn.: Greenwood), pp. 85–108.

Boli, J. and G. M. Thomas (1997), 'World Culture in the World Polity: A Century of Non-governmental Organization', *American Sociological Review*, **62:** 171–90.

Bond, E. J. (1996), *Ethics and Human Well-being* (Oxford: Blackwell).

Bracher, M. (1994), 'On the Psychological and Social Functions of Language'. In M. Bracher et al. (eds), *Lacanian Theory of Discourse* (New York: New York University Press), pp. 107–28.

Bramwell, A. (1989), *Ecology in the Twentieth Century* (New Haven, Conn.: Yale University Press).

———(1994), *The Fading of the Greens: The Decline of Environmental Politics in the West* (New Haven, Conn.: Yale University Press).

Breen, R. and D. Rottman (1998), 'Is the National State the Appropriate Geographical Unit for Class Analysis?', *Sociology*, **32:** 1–21.

Brilmayer, L. (1999), 'Realism Revisited'. In I. Shapiro and L. Brilmayer (eds), *Global Justice* (New York: New York University Press), pp. 192–216.

Brown, J. H. (1995), *Macroecology* (Chicago: University of Chicago Press).

Brownlee, D. (1998), 'Ancient Cosmic Spherules', *Nature*, **395:** 113–15.

Bull, H. (1977), *The Anarchical Society* (New York: Columbia University Press).

Bunting, R. (1997), *The Pacific Raincoast: Environment and Culture in an American Eden, 1778–1900* (Lawrence: University Press of Kansas).

Burhenne, W. (ed.) (1997), *International Environmental Law* (London: Kluwer).

Bwango, A., J. Wright, C. Elias and I. Burton (2000), 'Reconciling National and Global Priorities in Adaptation to Climate Change', *Environmental Monitoring and Assessment*, **61:** 145–59.

Caldwell, L. K. (1990), *International Environmental Policy*, 2nd edn (Durham, N.C.: Duke University Press).

Callicott, J. Baird (1994), *Earth's Insights* (Berkeley: University of California Press).

Campbell, R. (1996), 'Can Biology Make Ethics Objective?', *Biology and Philosophy*, **11**: 21–31.

Carlos, A. M. and F. D. Lewis (1999), 'Property Rights, Competition, and Depletion in the Eighteenth-century Canadian Fur Trade', *Canadian Journal of Economics*, **32**: 705–28.

Castells, M. (2000a), 'Materials for an Exploratory Theory of the Network Society', *British Journal of Sociology*, **51**: 5–24.

———(2000b), *The Information Age* (Oxford: Blackwell).

Cerny, P. G. (1996), 'Globalization and the Changing Logic of International Action', *International Organization*, **49**: 595–625.

Chanwai, K. and B. Richardson (1998), 'Re-working Indigenous Customary Rights?', *New Zealand Journal of Environmental Law*, **2**: 157–86.

Chase-Dunn, C., Y. Kawano and B. D. Brewer (2000), 'Trade Globalization since 1795: Waves of Integration in the World-system', *American Sociological Review*, **65**: 77–95.

Chechile, R. A. and S. Carlisle (eds) (1994), *Environmental Decision Making: A Multidisciplinary Perspective* (New York: Van Nostrand Reinhold).

Cheru, F. (1997), 'Global Apartheid and the Challenge to Civil Society'. In R. W. Cox (ed.), *The New Realism: Perspectives on Multilateralism and World Order* (London: Macmillan), pp. 205–22.

Chinn, C. (1995), *Poverty amidst Prosperity: The Urban Poor in England, 1834–1914* (Manchester: Manchester University Press).

Choucri, N. (ed.) (1993), *Global Accord: Environmental Challenges and International Responses* (Cambridge, Mass.: MIT Press).

Ciccantell, P. S. (1999), 'It's All About Power: The Political Economy and Ecology of Redefining the Brazilian Amazon', *Sociological Quarterly*, **40**: 293–316.

Cittadino, E. (1990), *Nature as the Laboratory: Darwinian Plant Ecology in the German Empire, 1880–1900* (Cambridge: Cambridge University Press).

———(1993), 'The Failed Promise of Human Ecology'. In M. Shortland (ed.), *Science and Nature: Essays in the History of the Environmental Sciences* (BSHS Monographs, 8): 251–84.

Clapp, J. (1998), 'The Privatization of Global Environmental Governance', *Global Governance*, **4**: 295–317.

Clark, A. M., E. J. Friedman and K. Hochstetler (1998), 'The Sovereign Limits of Global Civil Society: A Comparison of NGO Participation in UN World Conferences on the Environment, Human Rights, and Women', *World Politics*, **51**: 1–35.

Clark, N. (1997), 'Panic Ecology: Nature in the Age of Superconductivity', *Theory, Culture and Society*, **14**: 77–98.

Clarke, J. J. (1997), *Oriental Enlightenment: The Encounter between Asian and Western Thought* (London: Routledge).

Coate, S. and S. Morris (1999), 'Policy Persistence,' *American Economic Review*, **89**: 1327–36.

Coates, P. (1998), *Nature: Western Attitudes since Ancient Times* (Berkeley: University of California Press).

Cohen, J. (1996), *How Many People Can the Earth Support?* (New York: Norton).

Collins, R. (1998), *The Sociology of Philosophies: A Global Theory of Intellectual Change* (Cambridge, Mass.: Harvard University Press).

Cooper, D. E. (1999), 'Human Sentiment and the Future of Wildlife'. In F. L. Dolins (ed.), *Attitudes to Animals* (Cambridge: Cambridge University Press), pp. 231–43.

Costanza, R. et al. (1997), 'The Value of the World's Ecosystem Services and Natural Capital', *Nature*, **387:** 253–60.

Cox, R. W. (1997), 'Some Reflections on the Oslo Symposium'. In S. Gill (ed.), *Globalization, Democratization and Multilateralism* (London: Macmillan), pp. 245–52.

Cresswell, T. (1996), *In Place/Out of Place: Geography, Ideology, and Transgression* (Minneapolis: University of Minnesota Press).

Crook, S. (1998), 'Minotaurs and Other Monsters: "Everyday Life" in Recent Social Theory', *Sociology*, **32:** 523–40.

Crotty, J., G. A. Epstein and P. Kelly (1998), 'Multinational Corporations in the Neo-liberal Regime'. In D. Baker, G. A. Epstein and R. Pollin (eds), *Globalization and Progressive Economic Policy* (Cambridge: Cambridge University Press), pp. 117–43.

Crutzen, P. J. (1987), 'Role of the Tropics in Atmospheric Chemistry'. In R. E. Dickinson, (ed.), *The Geophysiology of Amazonia* (New York: Wiley).

Daily, G. (1997), *Nature's Services* (Washington, DC: Island Press).

Daily, G. and B. Walker (2000), 'Seeking the Great Transition', *Nature*, **403:** 243–5.

Daly, H. E. and K. N. Townsend (eds) (1992), *Valuing the Earth: Economics, Ecology, Ethics* (Cambridge, Mass.: MIT Press).

Dant, T. (1991), *Knowledge, Ideology and Discourse* (London: Routledge).

Darling, F. Fraser (1970), *Wilderness and Plenty* (London: BBC).

Dasgupta, S., H. Hettige and D. Wheeler (2000), 'What Improves Environmental Compliance? Evidence from Mexican Industry', *Journal of Environmental Economics and Management*, **39:** 39–66.

Davidson, J. (2000), 'Sustainable Development: Business as Usual or a New Way of Living?', *Environmental Ethics*, **22:** 25–42.

Davradou, M. and P. Wood (2000), 'The Promotion of Individual Autonomy in Environmental Ethics', *Environmental Ethics*, **22:** 73–84.

De Groot, W. T. (1992), *Environmental Science Theory* (London: Elsevier).

Deudney, D. (1990), 'The Case against Linking Environmental Degradation and National Security', *Millennium*, **19:** 461–76.

D'Hondt, S. (1998), 'Theories of Terrestrial Mass Extinction by Extraterrestrial Objects', *Earth Sciences History*, **17:** 157–73.

Dickinson, W. R. (2000), 'Effectively Responding to the Threat of Global Warming', *EOS*, **81:** 2.

Dobson, A. (1993), 'Critical Theory and Green Politics'. In A. Dobson and P. Lucardie (eds), *The Politics of Nature: Explorations in Green Political Theory* (London: Routledge), pp. 190–209.

Doern, G. B., L. A. Pal and B. W. Tomlin (1996), *Border Crossings: The Internationalization of Canadian Public Policy* (Toronto: Oxford University Press).

Dombrowski, P. (1998), 'Fragmenting Identities, Shifting Loyalties: The Influence of Individualization on Global Transformations', *Global Society*, **12:** 363–88.

Dowding, K. and D. King (1995), 'Introduction'. In K. Dowding and D. King (eds), *Preferences, Institutions, and Rational Choice* (Oxford: Clarendon Press), pp. 1–19.

Drayton, R. (2000), *Nature's Government: Science, British Imperialism and the 'Improvement' of the World* (New Haven, Conn.: Yale University Press).

Drury, S. (1999), *Stepping Stones: The Making of Our Home World* (Oxford: Oxford University Press).

Dunoff, J. L. (1999), 'The Death of the Trade Regime', *European Journal of International Law*, **10:** 733–62.

Dwivedi, O. P. (1986), 'Political Science and the Environment', *International Social Science Journal*, **38:** 377–90.

———(1994), *Environmental Ethics* (New Delhi: Sanchar).

Easterbrook, G. (1995), *A Moment on the Earth: The Coming Age of Environmental Optimism* (New York: Viking).

Economy, E. and M. A. Schreurs (1997), 'Domestic and International Linkages in Environmental Politics'. In M. A. Schreurs and E. Economy (eds), *The Internationalization of Environmental Protection* (Cambridge: Cambridge University Press), pp. 1–18.

Eder, K. (1996), *The Social Construction of Nature* (London: Sage).

EEA (European Environment Agency) (1998) *Europe's Environment: The Second Assessment* (Oxford: Elsevier).

Egri, C. (1999), 'Nature in Spiritual Traditions'. In F. Fischer and M. A. Hajer (eds), *Living with Nature: Environmental Politics as Cultural Discourse* (Oxford: Oxford University Press), pp. 58–79.

Ellis, R. D. (1998), *Just Results: Ethical Foundations for Policy Analysis* (Washington, DC: Georgetown University Press).

Emirbayer, M. and A. Mische (1998), 'What Is Agency?', *American Journal of Sociology*, **103:** 962–1023.

Ernst, W. G. (2000), *Earth Systems: Processes and Issues* (Cambridge: Cambridge University Press).

Faist, T. (1998), 'International Migration and Transnational Social Spaces', *European Journal of Sociology*, **39:** 213–47.

Farber, D. A. and C. Juma (1999), 'Eco-pragmatism: Making Sensible Environmental Decisions in an Uncertain World', *Nature*, **399:** 653.

Faucheux, S. and G. Froger (1995), 'Decision-making under Environmental Uncertainty', *Ecological Economics*, **15:** 29–42.

Ferleger, L. and J. R. Mandle (eds) (2000), 'Dimensions of Globalization', Special Issue, *The Annals of the American Academy of Political and Social Science*, **570**.

Finlayson, A. (1994), *Fishing for Truth: A Sociological Analysis of Northern Cod Stock Assessments from 1977 to 1990* (St John's: ISER).

Flew, A. (1991), *Thinking about Social Thinking*, 2nd edn (London: Fontana).

Fludernik, M. (1999), 'Cross-mirrorings of Alterity', *ARIEL*, **30:** 29–62.

Fogg, B. J. and C. Nass (1997), 'Silicon Sycophants: The Effects of Computers that Flatter', *International Journal of Human-Computer Studies*, **46:** 551–61.

Folke, C. and T. Kåberger (1991), 'Recent Trends in Linking the Natural Environment and the Economy'. In C. Folke and T. Kåberger (eds), *Linking the Natural Environment and the Economy* (Dordrecht: Kluwer), pp. 273–300.

Foltz, R. (2000), 'Is there an Islamic Environmentalism?', *Environmental Ethics*, **22:** 63–72.

Fomerand, J. (1996), 'UN Conferences: Media Events or Genuine Diplomacy?', *Global Governance*, **2:** 361–76.

Foster, J. B. (1999), 'Marx's Theory of Metabolic Rift: Classical Foundations for Environmental Sociology', *American Journal of Sociology*, **105:** 366–405.

Frank, D. J., A. Hironaka and E. Schofer (2000), 'The Nation-state and the Natural Environment over the Twentieth Century', *American Sociological Review*, **65:** 96–116.

Freeden, M. (1998), 'Is Nationalism a Distinct Ideology?', *Political Studies*, **46:** 748–65.

Frost, B.-P. (1999), 'A Critical Introduction to Alexandre Kojève's *Esquisse d'une phénoménologie du droit*', *Review of Metaphysics*, **52:** 595–640.

Furnes, H. and H. Standigel (1999), 'Biological Mediation in Ocean Crust Alteration', *Earth and Planetary Science Letters*, **166:** 97–103.

Gale, F. (1998), 'Constructing Global Civil Society Actors: An Anatomy of the Environmental Coalition Contesting the Tropical Timber Trade Regime', *Global Society*, **12:** 343–62.

Gasman, D. (1998), *Haeckel's Monism and the Birth of Fascist Ideology* (New York: Peter Lang).

Gaston, K. J. and T. M. Blackburn (1999), 'A Critique for Macroecology', *Oikos*, **84:** 353–68.

Gehrels, T. (1999), 'A Review of Comet and Asteroid Statistics', *Earth, Planets and Space*, **51:** 1155–61.

Gentry, B. S. (ed.) (1998), *Private Capital Flows and the Environment: Lessons from Latin America* (Cheltenham: Edward Elgar).

George, A. L. (1999), 'Knowledge for Statecraft', *Scandinavian Political Studies*, **22:** 89–97.

Gergen, K. J. (1999), *An Invitation to Social Construction* (London: Sage).

Gerring, J. (1999), 'What Makes a Concept Good?', *Polity*, **31:** 357–94.

Gibson, A. (1996), *Postmodern Theory* (Edinburgh: Edinburgh University Press).

Giddens, A. (1984), *The Constitution of Society: Outline of the Theory of Structuration* (Berkeley: University of California Press).

———(1987), 'Structuralism, Post-structuralism and the Production of Culture'. In A. Giddens and J. H. Turner (eds), *Social Theory Today* (Stanford, Calif.: Stanford University Press), pp. 195–221.

———(1989), 'A Reply to my Critics'. In D. Held and J. B. Thompson, *Social Theory of Modern Societies* (Cambridge: Cambridge University Press), pp. 249–301.

———(1991), *Modernity and Self-identity: Self and Society in the Late Modern Age* (Stanford, Calif.: Stanford University Press).

Gill, S. and D. Law (1988), *The Global Political Economy* (Baltimore: Johns Hopkins University Press).

Ginsberg, M. (1974), 'The Persistence of Individualism in the Theory of International Relations'. In R. Fletcher (ed.), *The Science of Society* (London: Heinemann), pp. 167–76.

Gismondi, M. (1997), 'Sociology and Environmental Impact Assessment', *Canadian Journal of Sociology*, **22:** 457–80.

Goldblatt, D. (1996), *Social Theory and the Environment* (Cambridge: Polity Press).

Goldstein-Gidoni, O. (1999), 'Kimono and the Construction of Gendered and Cultural Identities', *Ethnology*, **38:** 351–70.

Goldthorpe, J. H. (1998), 'Rational Action Theory for Sociology', *British Journal of Sociology*, **49:** 167–92.

Gollub, J. and M. Cross (2000), 'Chaos in Space and Time', *Nature*, **404:** 710–11.

Goodin, R. E. (1992), *Motivating Political Morality* (Oxford: Blackwell).

Gordon, L. R. (2000), 'Du Bois's Humanistic Philosophy of Human Sciences', *Annals of the American Academy of Political and Social Science*, **568:** 265–80.

Gorra, M. (1997), *After Empire: Scott, Naipaul, Rushdie* (Chicago: University of Chicago Press).

Grabel, I. (2000), 'The Political Economy of "Policy Credibility": The New-Classical Macroeconomics and the Remaking of Emerging Economies', *Cambridge Journal of Economics*, **24:** 1–19.

Gross, M. L. (1993), 'The Collective Dimensions of Political Morality', *Political Studies*, **42:** 40–61.

Grossman, L. S. (1998), *The Political Ecology of Bananas: Contract Farming, Peasants, and Agrarian Change in the Eastern Caribbean* (Chapel Hill: University of North Carolina Press).

Grove, J. W. (1999), 'The Face of Science at the End of the Twentieth Century', *Queen's Quarterly*, **106:** 383–91.

Grubb, M., with C. Vrolijk and D. Brack (1999), *The Kyoto Protocol* (London: RIIA).

Guinness, P. (1999), 'Local Community and the State', *Canberra Anthropology*, **22:** 88–110.

Gulati, R. and M. Gargiulo (1999), 'Where Do Interorganizational Networks Come from?', *American Journal of Sociology*, **104:** 1439–93.

Haas, E. B. (1990), *When Knowledge is Power: Three Models of Change in International Organizations* (Berkeley: University of California Press).

Habermas, J. (1987), *The Philosophical Discourse of Modernity*, tr. F. Lawrence (Cambridge, Mass.: MIT Press).

———(1989 [1962]), *The Structural Transformation of the Public Sphere*, tr. T. Burger (Cambridge, Mass.: MIT Press).

Haenn, N. (1999), 'The Power of Environmental Knowledge: Ethnoecology and Environmental Conflicts in Mexican Conservation', *Human Ecology*, **27:** 477–92.

Haila, Y. (1999), 'The North as/and the Other: Ecology, Domination, Solidarity'. In F. Fischer and M. A. Hajer (eds), *Living with Nature: Environmental Politics as Cultural Discourse* (Oxford: Oxford University Press), pp. 42–57.

Hamilton, R. F. (1996), *The Social Misconstruction of Reality* (New Haven, Conn.: Yale University Press).

Hannigan, J. A. (1995), *Environmental Sociology: A Social Constructionist Perspective* (London: Routledge).

Hannon, B. and M. Ruth (eds) (1997), *Modelling Dynamic Biological Systems* (New York: Springer).

Hansen, J. *et al.* (2000), 'Climate Modeling in the Global Warming Debate', in D. A. Randall (ed.), *General Circulation Model Development* (San Diego: Academic Press), pp. 127–64.

Harré, R., J. Brockmeier and P. Muhlhausler (1999), *Greenspeak: A Study of Environmental Discourse* (London: Sage).

Harrison, B. (1999), '"White Mythology" Revisited: Derrida and his Critics on Reason and Rhetoric', *Critical Inquiry*, **25:** 505–34.

Healy, K. (1998), 'Conceptualizing Constraint: Mouzelis, Archer and the Concept of Social Structure', *Sociology*, **32:** 509–22.

Hecht, A. D. (1999), 'The Triad of Sustainable Development', *Journal of Environment and Development*, **8:** 111–32.

Hedetoft, U. (1999), 'The Nation-state Meets the World: National Identities in the Context of Transnationality and Cultural Globalization', *European Journal of Social Theory*, **2:** 71–94.

Hedley, R. A. (1999), 'Transnational Corporations and Their Regulation', *International Journal of Comparative Sociology*, **40:** 215–30.

Held, D., A. McGrew, D. Goldblatt and J. Perraton (1999), *Global Transformations: Politics, Economics and Culture* (Stanford, Calif.: Stanford University Press).

Heller, A. (1990), 'Sociology as the Defetishization of Modernity'. In M. Albrow and E. King (eds), *Globalization, Knowledge and Society* (London: Sage), pp. 35–46.

Hensher, P. (2000), 'Silly but Not Like Us' (review), *Spectator*, 11 March: 34–5.

Higgins, R. W., J.-K. E. Schemm, W. Shi and A. Leetmaa (2000), 'Extreme Precipitation Events in the Western US related to Tropical Forcing', *Journal of Climate*, **13:** 793–820.

Hinich, M. J. and M. C. Munger (1997), *Analytical Politics* (Cambridge: Cambridge University Press).

Hirst, P. (1997), 'The Global Economy: Myths and Realities', *International Affairs*, **73.**

Holsti, K. J. (1995), *International Politics: A Framework for Analysis*, 7th edn (Englewood Cliffs, N.J.: Prentice-Hall).

Holton, R. J. (1996), 'Classical Social Theory'. In B. S. Turner (ed.), *The Blackwell Companion to Social Theory* (Oxford: Blackwell), pp. 25–52.

Houghton, J. T., L. G. Meira Filho, B. A. Callander, N. Harris, A. Kattenberg and K. Masbell (eds) (1996), *Climate Change 1995: The Science of Climate Change* (Cambridge: Cambridge University Press).

Houlahan, J. E., C. S. Findlay, B. R. Schmidt, A. H. Meyer and S. L. Kuzmin (2000), 'Quantitative Evidence for Global Amphibian Population Declines', *Nature*, **404:** 752–5.

Hungate, B. A. (1999), 'Ecosystem Responses to Rising Atmospheric CO_2'. In Yiqi Luo and H. A. Mooney (eds), *Carbon Dioxide and Environmental Stress* (San Diego, Calif.: Academic Press), pp. 265–85.

Hunt, L. (1990), 'History beyond Social Theory'. In D. Carroll (ed.), *The States of 'Theory'* (Stanford, Calif.: Stanford University Press), pp. 95–112.

Huntley, W. L. (1996), 'Kant's Third Image: Systemic Sources of the Liberal Peace', *International Studies Quarterly*, **40:** 45–76.

Iannaccone, L. R. (1997), 'Rational Choice: Framework for the Scientific Study of Religion'. In L. A. Young (ed.), *Rational Choice Theory and Religion* (London: Routledge), pp. 25–45.

Jameson, F. (1999), 'Notes on Globalization as a Philosophical Issue'. In F. Jameson and M. Miyoshi (eds), *The Cultures of Globalization* (Durham, N.C.: Duke University Press).

Jensen, L. A. (1997), 'Different Worldviews, Different Morals: America's Culture War Divide', *Human Development*, **40:** 325–44.

Jervis, R. (1997), *System Effects: Complexity in Political and Social Life* (Princeton, N.J.: Princeton University Press).

Johnson, M. (1990), 'NGOs at the Cross-roads in Indonesia'. In R. C. Rice (ed.), *Indonesian Economic Development* (Clayton, Victoria: Monash University, Centre of Southeast Asian Studies), pp. 77–92.

Jorgenson, J. P. and K. H. Redford (1993), 'Humans and Big Cats as Predators in the Neotropics', *Symposia of the Zoological Society of London*, **65:** 367–90.

Jørgensen, K. E. (2000), 'Continental IR Theory: The Best Kept Secret', *European Journal of International Relations*, **6:** 9–42.

Kabigumila, J. (1993), 'Feeding Habits of Elephants in Ngorongoro Crater, Tanzania', *African Journal of Ecology*, **31:** 156–63.

Keith, W. J. (1980), *The Poetry of Nature* (Toronto: University of Toronto Press).

Kershaw, A. P. and C. Whitlock (2000), 'Palaeoecological Records of the Last Glacial–Interglacial Cycle', *Palaeo*, **155:** 1–5.

Kiss, A. and D. Shelton (1997), *Manual of European Environmental Law*, 2nd edn (Cambridge: Cambridge University Press).

Knox, K. C. (1999), 'Lunatick Visions: Prophecy, Signs and Scientific Knowledge in 1790s London', *History of Science*, **37:** 427–58.

Kosso, P. (1998), *Appearance and Reality* (Oxford: Oxford University Press).

Kothari, R. (1993), 'The Yawning Vacuum: A World without Alternatives', *Alternatives*, **18:** 119–39.

Kovel, J. (1995), 'Ecological Marxism and Dialectic', *Capitalism, Nature, Socialism*, **6:** 31–50.

Krimsky, S. (2000), *Hormonal Chaos* (Baltimore: Johns Hopkins University Press).

Krips, H. (1987), *The Metaphysics of Quantum Theory* (Oxford: Clarendon Press).

Kroll-Smith, S., S. R. Couch and B. K. Marshall (1997), 'Sociology, Extreme Environments and Social Change', *Current Sociology*, **45:** 1–18.

Kula, E. (1998), *History of Environmental Economic Thought* (London: Routledge).

LaCapra, D. (1983), *Rethinking Intellectual History: Texts, Contexts, Language* (Ithaca, N.Y.: Cornell University Press).

Lachs, J. (1997), 'Valuational Species', *Review of Metaphysics*, **51:** 297–311.

Laferrière, E. and P. J. Stoett (1999), *International Relations Theory and Ecological Thought: Towards a Synthesis* (London: Routledge).

Langhelle, O. (1999), 'Sustainable Development: Exploring the Ethics of *Our Common Future*', *International Political Science Review*, **20:** 129–49.

Lapidge, M. (1988), 'The Stoic Inheritance'. In P. Dronke (ed.), *A History of Twelfth Century Western Philosophy* (Cambridge: Cambridge University Press), pp. 81–112.

Latour, B. (2000), 'When Things Strike Back', *British Journal of Sociology*, **51:** 107–23.

Law, D. (1997), 'Global Environmental Issues and the World Bank'. In S. Gill (ed.), *Globalization, Democratization and Multilateralism* (London: Macmillan), pp. 171–94.

Lawton, J. (1998), 'Small is Beautiful, and Very Strange', *Oikos*, **81:** 3–5.

———(1999), 'Are there General Laws in Ecology?', *Oikos*, **84:** 177–92.

Layder, D. (1997), *Modern Social Theory* (London: UCL Press).

Leiss, W. (ed.) (2001), *Risk Issue Management: A New Approach to Risk Controversies* (Montreal: McGill-Queen's University Press).

Lekan, T. (1999), 'Regionalism and the Politics of Landscape Preservation in the Third Reich', *Environmental History*, **4:** 384–404.

Le Prestre, P. (1997), *Ecopolitique Internationale* (Montreal: Guérin Universitaire).

Levin, R. (1993), 'The New Interdisciplinarity in Literary Criticism'. In N. Easterlin and B. Riebling (eds), *After Poststructuralism* (Evanston, Ill.: Northwestern University Press), pp. 13–44.

Levis, S., J. A. Foley and D. Pollard (2000), 'Large-scale Vegetation Feedbacks on a Doubled CO_2 Climate', *Journal of Climate*, **13:** 1313–25.

Lewontin, R. C. (1995), *Biology as Ideology* (Toronto: Anansi).

Light, A. and E. Katz (eds) (1996), *Environmental Pragmatism* (London: Routledge).

Livingston, I. (1997), *Arrow of Chaos: Romanticism and Postmodernity* (Minneapolis: University of Minnesota Press).

Low, S. M. (1997), 'Theorizing the City: Ethnicity, Gender and Globalization', *Critique of Anthropology*, **17:** 403–9.

Lowi, M. R. (1999), 'Water and Conflict in the Middle East and South Asia', *Journal of Environment and Development*, **8:** 376–96.

Lucardie, P. (1993), 'Why should Egocentrist become Ecocentrist?' In A. Dobson and P. Lucardie (eds), *The Politics of Nature: Explorations in Green Political Theory* (London: Routledge), pp. 21–38.

Ludwig, D. (1994), 'Missed Opportunities in Natural Resource Management', *Natural Resource Modelling*, **8:** 111–17.

Lyon, J. G. (2000), 'The Solar Wind–Magnetosphere–Ionosphere System', *Science*, **288:** 1987–91.

Lyotard, J.-F. (1993 [1989]), 'Oikos'. Republished in *Political Writings*, tr. B. Readings and K. Geiman (Minneapolis: University of Minnesota Press), pp. 96–107.

Macauley, R. (1994), *The Social Art: Language and its Uses* (Oxford: Oxford University Press).

McCarthy, F. (2000), 'What Does our Perspective on Deep Time Have to Offer in a New Millennium?', *Palaios*, **15:** 1–2.

McGrew, A. G. (1998), 'The Globalization Debate: Putting the Advanced Capitalist State in its Place', *Global Society*, **12:** 299–322.

McLaughlin, A. (1993), *Regarding Nature: Industrialism and Deep Ecology* (Albany, N.Y.: SUNY Press).

McMahon, M. (1997), 'From the Ground Up: Ecofeminism and Ecological Economics', *Ecological Economics*, **20:** 163–73.

McMichael, A. J. (1993), *Planetary Overload: Global Environmental Change and the Health of the Human Species* (Cambridge: Cambridge University Press).

Macnaghten, P. and J. Urry (1995), 'Towards a Sociology of Nature', *Sociology*, **29:** 203–20.

Macnaghten, P. and J. Urry (1998), *Contested Natures* (London: Sage).

McNeill, J. (1990), 'The Greening of International Relations', *International Journal*, **45:** 1–35.

Magdoff, H. (1992), 'Globalization – To What End?' In R. Miliband and L. Panitch (eds), *The Socialist Register 1992* (London: Merlin Press), pp. 44–75.

Magill-Evans, J., J. Darrah and M. Hodge (2000), 'Interdisciplinary Perspective on Linking Theory and Practice'. Paper to Tri-Joint Congress (CAOT/CPA/CASLPA), May.

Mapstone, E. R. (1996), 'Division of Labour in the Social Construction of Argument', *British Journal of Social Psychology*, **35:** 219–31.

Margulis, L. and J. E. Lovelock (1989), 'Gaia and Geognosy'. In M. B. Rambler et al. (eds), *Global Ecology* (New York: Academic Press), pp. 1–30.

Margulis, L. and L. Olendzenski, eds (1992), *Environmental Evolution: Effects of the Origin and Evolution of Life on Planet Earth* (Cambridge, Mass: MIT Press).

Marshall, G., ed. (1994), *The Concise Oxford Dictionary of Sociology* (Oxford: Oxford University Press).

Marsiliani, L. and T. Renström (2000), 'Time Inconsistency in Environmental Policy', *Economic Journal*, **110:** C123–8.

Martinez-Alier, J., with K. Schluepmann (1987), *Ecological Economics: Energy, Environment and Society* (Oxford: Blackwell).

Marx, K. (1973), *Grundrisse: Foundations of the Critique of Political Economy*, tr. M. Nicolaus (New York: Vintage).

Mathig, U. and E. de Mulder (1998), 'Report on the Geo-environmental Inquiry Project, Central and Eastern Europe', *Environmental Geology*, **35:** 37–40.

Meadowcroft, J. (1997), 'Planning, Democracy and the Challenge of Sustainable Development', *International Political Science Review*, **18:** 167–89.

Medin, D. L. and S. Atran (eds) (1999), *Folkbiology* (Cambridge, Mass.: MIT Press).

Meyer, J. W., J. Boli, G. M. Thomas and F. O. Ramirez (1997), 'World Society and the Nation-state', *American Journal of Sociology*, **103:** 144–81.

Midgley, M. (1997), 'Sustainability and Moral Pluralism'. In T. D. J. Chappell (ed.), *The Philosophy of the Environment* (Edinburgh: Edinburgh University Press), pp. 89–101.

Miller, R. B. (1996), *Casuistry and Modern Ethics: A Poetics of Practical Reasoning* (Chicago: University of Chicago Press).

Milton, K. (1996), *Environmentalism and Cultural Theory* (London: Routledge).

Minkoff, D. C. (1997), 'The Sequencing of Social Movements', *American Sociological Review*, **62:** 779–99.

Mitrany, D. (1933), *The Progress of International Government* (London: Allen and Unwin).

——— (1975), *The Functional Theory of Politics* (London: Martin Robertson).

Mittelman, J. H. (1994), 'The Globalization Challenge: Surviving at the Margins', *Third World Quarterly*, **15.**

Mittelman, J. H. and M. K. Pasha (1997), *Out from Underdevelopment Revisited* (London: Macmillan).

Moore, B. (1920), 'The Scope of Ecology', *Ecology*, **1:** 3–5.

Morrison, P. (1996), *The Poetics of Fascism: Ezra Pound, T. S. Eliot, Paul de Man* (Oxford: Oxford University Press).

Murphy, R. (1995), 'Sociology as if Nature did not Matter: An Ecological Critique', *British Journal of Sociology*, **46:** 667–707.

Myerson, G. and Y. Rydin (1996), *The Language of Environment* (London: UCL Press).

Naess, A. (1999), 'The Principle of Intensity', *Journal of Value Inquiry*, **33:** 5–9.

Nederveen Pieterse, J. (2000a), 'Globalization North and South', *Theory, Culture and Society*, **17:** 129–37.

——— (ed.) (2000b), *Global Futures: Shaping Globalization* (London: Zed).

Needham, J. (1968 [1936]), *Order and Life* (Cambridge, Mass.: MIT Press).

Neimanis, V. and A. Kerr (1996), 'Developing National Environmental Indicators'. In A. R. Berger and W. J. Iams (eds), *Geoindicators* (Rotterdam: Balkema), pp. 369–76.

Nelson, C. (1994), 'Care in Feeding: Vegetarianism and Social Reform in Alcott's America'. In C. Nelson and L. Vallone (eds), *The Girl's Own: Cultural Histories of the Anglo-American Girl, 1830–1915* (Athens: University of Georgia Press), pp. 11–33.

Nelson, R. H. (1997), 'In Memoriam: On the Death of the "Market Mechanism"', *Ecological Economics*, **20:** 187–97.

Nicholson, M. (1995), 'Rational Decision in International Crises'. In K. Dowding and D. King (eds), *Preferences, Institutions, and Rational Choice* (Oxford: Clarendon Press).

Noorduyn, R. E. and W. T. De Groot (1999), 'Environment and Security: Improving the Interaction of Two Science Fields', *Journal of Environment and Development*, **8:** 24–48.

Norcross, A. (1998), 'Speed Limits, Human Lives, and Convenience', *Philosophy and Public Affairs*, **27:** 59–64.

Norton, B. G. (1998), 'Improving Ecological Communication', *Ecological Applications*, **8:** 350–64.

Noss, R. (1999), 'Is there a Special Conservation Biology?', *Ecography*, **22:** 113–22.

O'Connor, J. (1998), *Natural Causes: Essays in Ecological Marxism* (New York: Guilford Press).

O'Hara, S. (1999), 'Economics, Ecology, and Quality of Life: Who Evaluates?', *Feminist Economics*, **5:** 83–9.

O'Riordan, T. (1994), *Environmental Science for Environmental Management* (London: Wiley).

Paehlke, R. (1989), *Environmentalism and the Future of Progressive Politics* (New Haven, Conn.: Yale University Press).

——— (1996), 'Environmental Challenges to Democratic Practice'. In W. M. Lafferty and J. Meadowcroft (eds), *Democracy and the Environment: Problems and Prospects* (Cheltenham: Edward Elgar), pp. 18–38.

Page, E. (1999), 'Intergenerational Justice and Climate Change', *Political Studies*, **47**: 53–66.

Pahl-Wostl, C. (1995), *The Dynamic Nature of Ecosystems: Chaos and Order Entwined* (New York: Wiley).

Parsons, T. (1951), *The Social System* (New York: Free Press).

Patten, B. C. (1991), 'Network Ecology'. In M. Higashi and T. P. Burns (eds), *Theoretical Studies of Ecosystems: The Network Perspective* (Cambridge: Cambridge University Press), pp. 288–351.

Paul, R. A. (2000), 'Sons or Sonnets: Nature and Culture in a Shakespearean Anthropology', *Current Anthropology*, **41**: 1–10.

Pearson, G. A. (1924), 'Some Conditions for Effective Research', *Science*, **60**: 71–3.

Pels, D. (1995), 'Knowledge Politics and Anti-politics', *Theory and Society*, **24**: 79–104.

Peters, R. H. (1991), *A Critique for Ecology* (Cambridge: Cambridge University Press).

Peterson, M. J. (1998), 'Organizing for Effective Environmental Cooperation', *Global Governance*, **4**: 415–38.

Pimm, S. L. (1991), *The Balance of Nature? Ecological Issues in the Conservation of Species and Communities* (Chicago: University of Chicago Press).

Platt, J. and S. Hopper (1997), 'Fragmentation, Social Theory, and Feminism', *Contemporary Sociology*, **26**: 283–5.

Polis, G. A. (1998), 'Stability is Woven by Complex Webs', *Nature*, **395**: 744–5.

Portes, A. (2000), 'The Hidden Abode', *American Sociological Review*, **65**: 1–18.

Pouyat, R. V. (1999), 'Science and Environmental Policy', *BioScience*, **49**: 281–6.

Prakash, A. and J. A. Hart (2000), 'Indicators of Economic Integration', *Global Governance*, **6**: 95–114.

Princen, T. (1998), 'From Property Regime to International Regime: An Ecosystems Perspective', *Global Governance*, **4**: 395–414.

Provencal, V. (1997), 'Writing the Large Letter Small: The Analogy of State and Individual at *Republic V 462c–d*', *Apeiron*, **30**: 73–88.

Quesnel, L., ed. (1995), *Social Sciences and the Environment* (Ottawa: University of Ottawa Press).

Raustiala, K. (1997), 'States, NGOs, and International Environmental Institutions', *International Studies Quarterly*, **41**: 719–40.

Rawls, J. (1999), *The Law of Peoples* (Cambridge, Mass.: Harvard University Press).

Redclift, M. and T. Benton, eds (1994), *Social Theory and the Global Environment* (London: Routledge).

Redclift, M. and G. Woodgate (eds) (1997), *The International Handbook of Environmental Sociology* (Cheltenham: Edward Elgar).

Rice, T. J. (1997), *Joyce, Chaos and Complexity* (Urbana, Ill.: University of Illinois Press).

Riddell-Dixon, E. (1999), 'Mainstreaming Women's Rights', *Global Governance*, **5**: 149–72.

Risse, T. (2000), '"Let's Argue!": Communicative Action in World Politics', *International Organization*, **54**: 1–39.

Robertson, R. (1995), 'Glocalization: Time-Space and Homogeneity-Heterogeneity'. In M. Featherstone et al. (eds), *Global Modernities* (London: Sage).
Rochefort, D. A. and R. W. Cobb, eds (1994), *The Politics of Problem Definition: Shaping the Policy Agenda* (Lawrence: University Press of Kansas).
Rolston, H. (1988), *Environmental Ethics* (Philadelphia: Temple University Press).
———(1993), 'God and Endangered Species'. In L. S. Hamilton (ed.), *Ethics, Religion and Biodiversity* (Cambridge, Mass.: White Horse Press), pp. 40–64.
Rorty, R. (1989), *Contingency, Irony, and Solidarity* (Cambridge: Cambridge University Press).
Rosenau, J. N. and M. Durfee (1995), *Thinking Theory Thoroughly: Coherent Approaches to an Incoherent World* (Boulder, Colo.: Westview).
Rovane, C. (1998), *The Bounds of Agency: An Essay in Revisionary Metaphysics* (Princeton, N.J.: Princeton University Press).
Rowlands, I. H. (1995), *The Politics of Global Atmospheric Change* (Manchester: Manchester University Press).
Russell, A. and R. Dennis (2000), 'NARSTO Critical Review of Photochemical Models and Modeling', *Atmospheric Environment*, **34:** 2283–324.
Saarnio, S., T. Saarinen, H. Vasander and J. Silvola (2000), 'A Moderate Increase in the Annual CH_4 Efflux by Raised CO_2 and NH_4NO_3 Supply in a Boreal Oligotrophic Mire', *Global Change Biology*, **6:** 137–44.
Sabini, J. and J. Schulkin (1994), 'Biological Realism and Social Constructionism', *Journal for the Theory of Social Behaviour*, **24:** 207–18.
Sackman, D. C. (2000), '"Nature's Workshop": The Work Environment and Workers' Bodies in California's Citrus Industry, 1900–1940', *Environmental History*, **5:** 27–53.
Said, E. (2000), 'Invention, Memory, and Place', *Critical Inquiry*, **26:** 175–92.
Salter, L. and A. Hearn, eds (1996), *Outside the Lines: Issues in Interdisciplinary Research* (Montreal: McGill-Queen's University Press).
Sandler, T. (1997), *Global Challenges: An Approach to Environmental, Political and Economic Problems* (Cambridge: Cambridge University Press).
Sargisson, L. (1996), *Contemporary Feminist Utopianism* (London: Routledge).
Saul, J. (1992), *Voltaire's Bastards* (Toronto: Viking).
Schellnhuber, H. J. (1999), '"Earth System" Analysis and the Second Copernican Revolution', *Nature*, **402:** C19–23.
Schellnhuber, H. J. and V. Wenzel, eds (1998), *Earth System Analysis* (New York: Springer).
Schlesinger, W. H. (1991), *Biogeochemistry: An Analysis of Global Change* (San Diego, Calif.: Academic Press).
———(1997), *Biogeochemistry: An Analysis of Global Change*, 2nd edn (San Diego, Calif.: Academic Press).
Schmidt, B. C. (1998), *The Political Discourse of Anarchy: A Disciplinary History of International Relations* (Albany, N.Y.: SUNY Press).
Schoijet, M. (1999), '*Limits to Growth* and the Rise of Catastrophism', *Environmental History*, **4:** 515–30.
Schokkaert, E. and J. Eyckmans (1999), 'Greenhouse Negotiations and the Mirage of Partial Justice'. In M. H. I. Dore and T. D. Mount (eds), *Global Environmental Economics* (Oxford: Blackwell), pp. 193–217.
Schoonmaker, D. (1997), 'El Niño and Unemployment', *American Scientist*, **85:** 319–21.

Scriven, T. (1997), *Wrongness, Wisdom, and Wilderness: Toward a Libertarian Theory of Ethics and the Environment* (Albany, N.Y.: SUNY Press).

Simpson, E. (1999), 'Between Internalism and Externalism in Ethics', *Philosophical Quarterly*, **49:** 201–14.

Singer, S. F. (1998), 'Scientific Case against the Global Climate Treaty', *Environmental Geology*, **35:** 287–9.

Singer, T. O. and R. Stumberg (1999), 'A Multilateral Agreement on Investment', *Journal of Environment and Development*, **8:** 5–23.

Sjolander, C. T. (1996), 'The Rhetoric of Globalization', *International Journal*, **51:** 603–16.

Sjöstedt, G. (1993), 'Special and Typical Attributes of International Environmental Negotiations'. In G. Sjöstedt et al. (eds), *International Environmental Negotiations* (Stockholm: Utrikespolitiska Institutet), pp. 22–39.

Smith, P. (1994), '*Mansfield Park* and the World Stage', *Cambridge Quarterly*, **23:** 203–29.

Smith, R. (1998), 'CanLit Takes it to the Street' (review), *Globe and Mail* (Toronto), 23 May: D10.

Smith, Z. A. (1995), *The Environmental Policy Paradox*, 2nd edn (Englewood Cliffs, N.J.: Prentice-Hall).

Sölderholm, P. (1999), 'Pollution Charges in a Transition Economy: The Case of Russia', *Journal of Economic Issues*, **33:** 403–10.

Sornig, K. (1989), 'Some Remarks on Linguistic Strategies of Persuasion'. In R. Wodak (ed.), *Language, Power and Ideology* (Amsterdam: John Benjamins), pp. 95–114.

Sprinz, D. (1994), 'Strategies of Inquiry into International Environmental Policy', *International Studies Notes*, **19:** 32–4.

Sprout, H. and M. Sprout (1962), *Foundations of International Politics* (New York: Van Nostrand).

———(1971), *Toward a Politics of the Planet Earth* (New York: Van Nostrand).

Spybey, T. (1996), *Globalization and World Society* (Cambridge: Polity Press).

Sterling-Folker, J. (2000), 'Competing Paradigms or Birds of a Feather? Constructivism and Neoliberal Institutionalism Compared', *International Studies Quarterly*, **44:** 97–119.

Stoett, P. J. (1995), 'Environmental Problems, Policies, and Prospects in Africa'. In O. P. Dwivedi and D. K. Vajpeyi (eds), *Environmental Policies in the Third World* (Westport, Conn.: Greenwood), pp. 109–24.

Sturgeon, N. (1999), 'Ecofeminist Appropriations and Transnational Environmentalisms', *Identities*, **6:** 255–80.

Sugiura, K. (1999), 'Institutional Economics Needs Inter–disciplinary Studies of Social Sciences', *Journal of Economic Issues*, **33:** 257–64.

Swirsky, P. (2000), *Between Literature and Science: Poe, Lem, and Explorations in Aesthetics, Cognitive Science, and Literary Knowledge* (Montreal: McGill–Queen's University Press).

Thiele, L. P. (1999), 'Evolutionary Narratives and Ecological Ethics', *Political Theory*, **27:** 6–38.

Thomas, J. D. and R. E. Dodge (1999), 'Quick Action Needed for World's Declining Coral Reefs', *Earth System Monitor*, **10:** 12–13, 16.

Thompson, J. B. (1989), 'The Theory of Structuration'. In D. Held and J. B. Thompson (eds), *Social Theory of Modern Societies* (Cambridge: Cambridge University Press), pp. 56–76.

Torfing, J. (1998), *Politics, Regulation and the Modern Welfare State* (London: Macmillan).

Touraine, A. (1995), *Critique of Modernity*, tr. D. Macey (Oxford: Blackwell).

Tranter, B. (1999), 'Environmentalism in Australia: Elites and the Public', *Journal of Sociology*, **35:** 331–50.

Tronto, J. C. (1996), 'Care as a Political Concept'. In N. J. Hirschmann and C. Di Stefano (eds), *Revisioning the Political: Feminist Reconstructions of Traditional Concepts in Western Political Theory* (Boulder, Colo.: Westview), pp. 139–56.

Tuler, P. (1996), *Meanings, Understandings, and Interpersonal Relationships in Environmental Policy Discourse*, Clark University: unpublished PhD dissertation.

Turner, H. (1999), 'Human Agency and Impersonal Determinants in Historical Causation', *History and Theory*, **38:** 300–6.

Ulanowicz, R. E. (2000), 'Quantifying Constraints upon Trophic and Migratory Transfers Landscapes'. In J. Sanderson and L. D. Harris (eds), *Landscape Ecology* (Boca Raton, Fla.: Lewis), pp. 113–42.

VanAsselt, M. B. A. and J. Rotmans (1996), 'Uncertainty in Perspective', *Global Environmental Change*, **6:** 121–58.

Vautier, M. (1998), *New World Myth: Postmodernism and Postcolonialism in Canadian Fiction* (Montreal: McGill-Queen's University Press).

Vayda, A. P. and B. B. Walters (1999), 'Against Political Ecology', *Human Ecology*, **27:** 167–80.

Vercelli, A. (1999), 'Environmental Uncertainty and Future Generations'. In M. H. I. Dore and T. D. Mount (eds), *Global Environmental Economics* (Oxford: Blackwell), pp. 77–92.

Verseghy, D. (2000), 'The Canadian Land Surface Scheme (CLASS)', *Atmosphere-Ocean*, **38:** 1–13.

Vitousek, P. M., et al. (1997), 'Human Alteration of the Global Nitrogen Cycle', *Ecological Applications*, **7:** 737–50.

Vogel, S. (1997), 'Habermas and the Ethics of Nature'. In R. S. Gottlieb (ed.), *The Ecological Community* (London: Routledge), pp. 175–92.

von Maltzahn, K. (1994), *Nature as Landscape: Dwelling and Understanding* (Montreal: McGill-Queen's University Press).

Voss, J. F. (1988), 'Problem Solving and Reasoning in Ill-structured Domains'. In C. Antaki (ed.), *Analyzing Everyday Explanation* (Beverly Hills, Calif.: Sage), pp. 74–93.

Wagner, A. (1999), 'Causality in Complex Systems', *Biology and Philosophy*, **14:** 83–101.

Wagner, S. A. (1997), *Understanding Green Consumer Behaviour* (London: Routledge).

Wagner, W., J. Valencia and F. Elejabarrieta (1996), 'Relevance, Discourses and the "Hot" Stable Core of Social Representations', *British Journal of Social Psychology*, **35:** 331–51.

Walker, B. (1998), 'Thoreau's Alternative Economics', *American Political Science Review*, **92:** 845–56.

Wallerstein, I. (1999), 'The Heritage of Sociology, the Promise of Social Science', *Current Sociology*, **47:** 1–43.

——(2000), 'From Sociology to Historical Social Science: Prospects and Obstacles', *British Journal of Sociology*, **51:** 25–36.

Wang, G. and E. Eltahir (2000), 'Ecosystem Dynamics and the Sahel Drought', *Geophysical Research Letters*, **27:** 795–8.

Wapner, P. (1996), *Environmental Activism and World Civic Politics* (Albany, N.Y.: SUNY Press).
——(1997), 'Environmental Ethics and Global Governance: Engaging the International Liberal Tradition', *Global Governance*, **3:** 213–31.
Ward, G. (1996), *Theology and Contemporary Critical Theory* (London: Macmillan).
Waterhouse, R. (1999), 'The Vision Splendid: Conceptualizing the Bush, 1813–1913', *Journal of Popular Culture*, **33:** 23–34.
Weber, T. P. (1999), 'A Plea for a Diversity of Scientific Styles in Ecology', *Oikos*, **84:** 526–9.
Weiner, D. R. (1992), 'Demythologizing Environmentalism', *Journal of the History of Biology*, **25:** 385–412.
Weinert, F. (1999), 'Theories, Models and Constraints', *Studies in History and Philosophy of Science*, **30A:** 303–33.
Weiss, L. (1998), *The Myth of the Powerless State* (Ithaca, N.Y.: Cornell University Press).
Weiss, E. B. and H. K. Jacobson, eds (1998), *Engaging Countries: Strengthening Compliance with International Environmental Accords* (Cambridge, Mass.: MIT Press).
Weix, G. G. (1998), 'Islamic Prayer Groups in Indonesia: Local Forums and Gendered Responses', *Critique of Anthropology*, **18:** 405–20.
Westing, A. H. (1999), 'Towards a Universal Recognition of Environmental Responsibilities', *Environmental Conservation*, **26:** 157–8.
Whittier, N. (1997), 'Political Generations, Micro-cohorts, and the Transformation of Social Movements', *American Sociological Review*, **62:** 760–78.
Wildavsky, A. (1989), *Craftways: On the Organization of Scholarly Work* (New Brunswick, N.J.: Transaction).
Williams, F. (1999), 'Good-enough Principles for Welfare', *Journal of Social Policy*, **28:** 667–87.
Williams, J. E. (2000), 'The Biodiversity Crisis and Adaptation to Climate Change', *Environmental Monitoring and Assessment*, **61:** 65–74.
Wilson, C. (1993), 'On Some Alleged Limitations to Moral Endeavor', *Journal of Philosophy*, **15:** 275–89.
Wilson, E. (2000), 'Emerson's *Nature*, Paralogy, and the Physics of the Sublime', *Mosaic*, **33:** 39–58.
Wilson, E. O. (1998), *Consilience: The Unity of Knowledge* (New York: Knopf).
Wilson, J. (1992), 'Green Lobbies: Pressure Groups and Environmental Policy'. In R. Boardman (ed.), *Canadian Environmental Policy: Ecosystems, Politics and Process* (Toronto: Oxford University Press), pp. 109–25.
——(1998), *Talk and Log: Wilderness Politics in British Columbia, 1965–96* (Vancouver: UBC Press).
Wolfe, A. (1993), *The Human Difference* (Berkeley: University of California Press).
Woods, N. (1999), 'Good Governance in International Organizations', *Global Governance*, **5:** 39–62.
World Bank (1994), *Indonesia: Environment and Development* (Washington, DC).
Young, O. R. (1995), *Global Governance: Drawing Insights from the Environmental Experience* (Hanover: Dartmouth College, Dickey Center).
Zafirovski, M. (1999), 'What is Really Rational Choice?', *Current Sociology*, **47:** 47–113.
Zürn, M. (1998), 'The Rise of International Environmental Politics: A Review of Current Research', *World Politics*, **50:** 617–49.

Index